W9-ANH-120

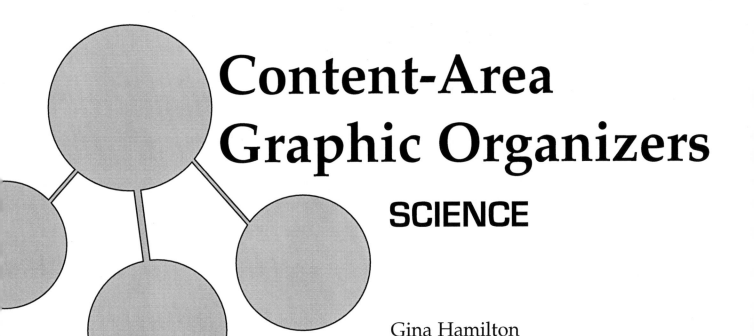

# Content-Area
# Graphic Organizers
## SCIENCE

Gina Hamilton

No Longer
the Property of
Bluffton University

Bluffton University Library

WALCH PUBLISHING

The classroom teacher may reproduce materials in this book for classroom use only.
The reproduction of any part for an entire school or school system is strictly prohibited.
No part of this publication may be transmitted, stored, or recorded in any form
without written permission from the publisher.

1    2    3    4    5    6    7    8    9    10

ISBN 0-8251-5037-X

Copyright © 2005

J. Weston Walch, Publisher

P. O. Box 658 • Portland, Maine 04104-0658

walch.com

Printed in the United States of America

# Table of Contents

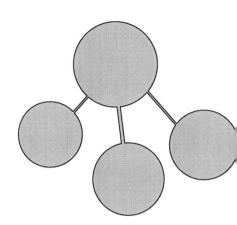

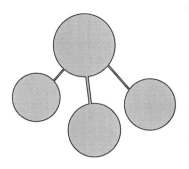

# To the Teacher

Graphic organizers can be a versatile tool in your classroom. Organizers offer an easy, straightforward way to visually present a wide range of material. Research suggests that graphic organizers support learning in the classroom for all levels of learners. Gifted students, students on grade level, and students with learning difficulties all benefit from their use. Graphic organizers reduce the cognitive demand on students by helping them access information quickly and clearly. Using graphic organizers, learners can understand content more clearly and can take clear, concise notes. Ultimately, learners find it easier to retain and apply what they've learned.

Graphic organizers help foster higher-level thinking skills. They help students identify main ideas and details in their reading. They make it easier for students to see patterns such as cause and effect, comparing and contrasting, and chronological order. Organizers also help learners master critical-thinking skills by asking them to recall, evaluate, synthesize, analyze, and apply what they've learned. Research suggests that graphic organizers contribute to better test scores because they help students understand relationships between key ideas and enable them to be more focused as they study.

This book shows students how they can use some common graphic organizers as they read and write in science classes. As they become familiar with graphic organizers, they will be able to adapt them to suit their needs.

In the science classroom, graphic organizers help students:
- preview new material
- make connections between new material and prior learning
- recognize patterns and main ideas in reading
- understand the relationships between key ideas
- organize information and take notes
- review material

This book offers graphic organizers suitable for science tasks, grouped according to big-picture skills, such as organizing knowledge and data; classifying information; comparing and contrasting; showing cause and effect; doing labs; and sequencing. Each organizer is introduced with an explanation of its primary uses and structure. Next comes a step-by-step description of how to create the organizer, with a worked-out example that uses text relevant to the content area. Finally, an application section asks students to use the techniques they have just learned to complete a blank organizer with information from a sample text. Throughout, learners are encouraged to customize the organizers to suit their needs. To emphasize the variety of graphic organizers available, an additional organizer suitable for each big-picture skill is introduced briefly at the end of each lesson.

*Content-Area Graphic Organizers: Science* is easy to use. Simply photocopy and distribute the section on each graphic organizer. Blank copies of the graphic organizers are included at the back of this book so that you can copy them as often as needed. The blank organizers are also available for download at our web site, walch.com.

As learners become familiar with using graphic organizers, they will develop their own approaches and create their own organizers. Encourage them to adapt them, change them, and create their own for more complex strategies and connections.

Remember, there is no one right way to use graphic organizers; the best way is the way that works for each student.

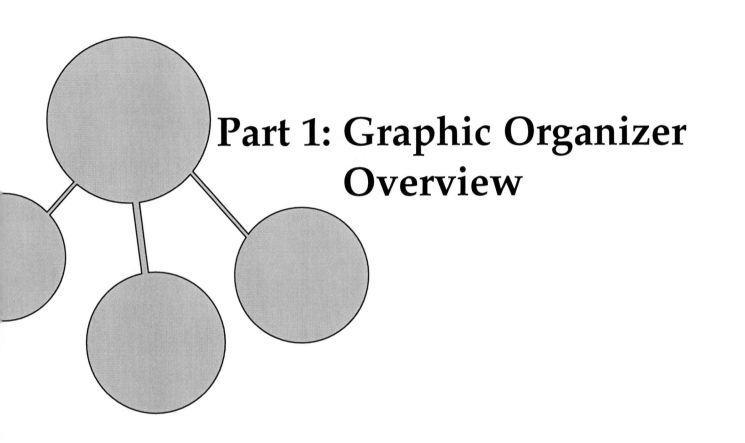

# Part 1: Graphic Organizer Overview

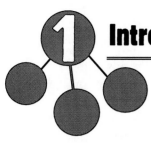

# Introduction to Graphic Organizers

You've probably heard the old saying, "A picture is worth a thousand words." Like most old sayings, it isn't always true. But in many things we do, words alone are not the best way to communicate. That's why we use pictures and, in particular, graphic organizers.

A graphic organizer is simply a special drawing that contains words or numbers. If you've ever made a web or filled in a chart, then you already know how to use a graphic organizer. In this book, you'll find that you can use graphic organizers in ways you may not have expected. And you'll find that they can make your learning a lot easier!

The power of a graphic organizer is that instead of just telling you about relationships among things, it can show them to you. A graphic organizer can help you understand information much more easily than the same information written out as a paragraph of text. For example, look at this listing of names, addresses, and telephone numbers. Use it to find the telephone number for Amanda Jones.

Alden E. Jones, 18 Milford St., Boston, MA 02118, (617) 555-8040. Alun Huw Jones, 91 Westland Ave., Boston, MA 02115, (617) 555-9654. Alvin Jones, 715 Tremont St., Boston, MA 02118, (617) 555-2856. Alvin D. Jones, 77 Salem St., Boston, MA 02113, (617) 555-2890. Amanda Jones, 111 W. 8th St., Boston, MA 02127, (617) 555-0738. Amos K. Jones, 11 Helen St., Boston, MA 02124, (617) 555-3560. Andre N. Jones, 523 Mass. Ave., Boston, MA 02118, (617) 555-0829. Andrew Jones, 168 Northampton St., Boston, MA 02118, (617) 555-0069.

In order to find Amanda's number you had to read, or at least scan, the whole text. Here is the same information presented in a graphic organizer—a table.

| Name | Address | City, State, Zip | Phone |
|------|---------|------------------|-------|
| Alden E. Jones | 18 Milford St. | Boston, MA 02118 | (617) 555-8040 |
| Alun Huw Jones | 91 Westland Ave. | Boston, MA 02115 | (617) 555-9654 |
| Alvin Jones | 715 Tremont St. | Boston, MA 02118 | (617) 555-2856 |
| Alvin D. Jones | 77 Salem St. | Boston, MA 02113 | (617) 555-2890 |
| Amanda Jones | 111 W 8th St. | Boston, MA 02127 | (617) 555-0738 |
| Amos K. Jones | 11 Helen St. | Boston, MA 02124 | (617) 555-3560 |
| Andre N. Jones | 523 Mass Ave. | Boston, MA 02118 | (617) 555-0829 |
| Andrew Jones | 168 Northampton St. | Boston, MA 02118 | (617) 555-0069 |

Which arrangement was easier to use? Most people find it easier to see the information in the table. This is because the table gives all the names in one column, all the telephone numbers in another column, and all the information about each person in one row. As soon as you know how the table is set up—the labels at the top of each column tell you—you can quickly find what you're looking for.

Graphic organizers use lines, circles, grids, charts, tree diagrams, symbols, and other visual elements to show relationships—classifications, comparisons, contrasts, time sequence, parts of a whole, and so on—much more directly than text alone.

You can use graphic organizers in many ways. You can use them before you begin a lesson to lay the foundation for new ideas. They can help you recall what you already know about a subject and see how new material is connected to what you already know.

You can use them when you are reading to take notes or to keep track of what you read. It doesn't matter what you are reading—a textbook, a biography, or an informational article. Organizers can help you understand and analyze what you read. You can use them to recognize patterns in the reading. They can help you identify the main idea and its supporting details. They can help you compare and contrast all kinds of things, from people to ideas, animals, and events.

Graphic organizers can help you after you read. You can use them to organize your notes and figure out the most important points in what you read. They are a great tool as you review to make sure you understood everything or to prepare for a test.

You can use graphic organizers when you write, too. They are particularly useful for prewriting and planning. Organizers can help you brainstorm new ideas. They can help you sort out the key points you want to make. Graphic organizers can help you write clearly and precisely.

Think of graphic organizers as a new language. Using this new language may be a bit awkward at first, but once you gain some fluency, you will enjoy communicating in a new way.

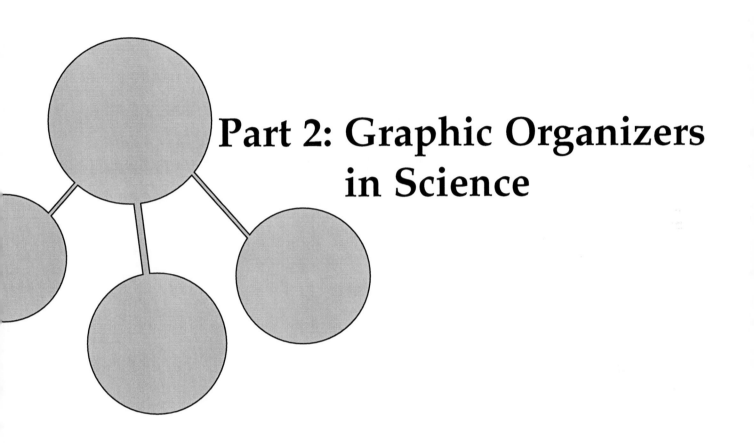

# Part 2: Graphic Organizers in Science

# Organizing Knowledge and Data

In every area of science, information must be collected and organized. As a student in a science class, you may have to collect and organize information from many sources. Here are the most common:

• Collecting material from books, the Internet, magazines, and journals to write research papers

• Taking notes from observations and organizing them to analyze the data

• Collecting information from multiple sources and combining it into one organized form

• Collecting information from experiments and using it to write lab reports

• Organizing information from surveys

• Organizing information from brainstorming sessions

When you collect information from a source, you need to organize it and be able to recall it. Graphic organizers can help. In this lesson, we will take an in-depth look at two important graphic organizers that can help you organize data and knowledge—a web and a table. We will also take a quick look at another method you can use for brainstorming problems and answers.

Until it is organized, raw data is virtually useless. The reason for this is that there is no true context in raw data. Humans in science, as in every other content area, place meaning on the data they collect.

Here is an example. Let's say that you collect rocks. Each individual rock is beautiful and has specific qualities, such as mass, color, and texture. However, until you organize the rocks in some way, the collection has no meaning. If you group the rocks by categories—igneous, sedimentary, and metamorphic—the collection makes some sense. You might choose to group them in some other way—by mineral content, for instance. The order that you impose has to make sense for the way you want to use the material.

The same thing is true of facts and information. Organizing information and data is the first step to making sense of the material. After it is organized in some way, you can do the harder job of classification.

**Webs** How you choose to organize information depends largely on what kind of information it is. Some information lends itself more easily to webs, some to charts, and some to other sorting methods.

Webs are graphic organizers that can show how facts or ideas are related. Webs don't just show the information; they show the overall structure of the information and the connections between different pieces of information. Webs can help you associate ideas and make connections you might not otherwise have made. They can also help you group information into categories. You can use webs to summarize an article or a chapter in a book. They are also useful in the early stages of writing an essay, as they can help you organize your material and see how ideas fit together.

Webs are particularly useful when you have information that relates to a main topic but does not fit into a hierarchy.

Webs are also very useful when you are combining data from more than one person. For example, if you are partnered with another student, or grouped together into a field study team, at some point, you have to combine and evaluate the group's data.

**Using Webs** Use webs when you have a lot of data that can be organized into multiple categories. This is a common feature in science. You can use lines to draw the relationships. One piece of information might relate to several categories. Making a web is a free-flowing process, allowing you to illustrate multiple relationships.

There are many ways to create a web. For each approach, you will need to identify categories that fit your information. You will also define the relationships. This imposition of order need make sense only to you.

In a web, you write the main idea in a circle in the middle of a page. If you know what subtopics you want to consider, write them in circles around the center. Then write all your facts and information about the topic in circles around the subtopics. When you see relationships between different pieces of information, you can draw lines to connect them.

Once all the information has been recorded you can look at the web to see how the pieces of information fit together. Do the related pieces of information suggest categories that you could use to organize the material? Are there any patterns in the web? Do you need any more information to make the pattern complete? If so, what more do you need to find out? Here are some things to keep in mind as you create a web:

• Each piece of information should have its own space on the web.

• Pieces of information that relate to a larger category should be linked to that category. (You might have an idea of the categories you will be watching for ahead of time, or you might not. Often, during an observation, new relationships and links become clear.)

- Check to see where data points can be cross-referenced. For instance, a single data point might relate to more than one category.

**Webs in Action**   Here are the notes two students made as they observed monarch butterflies in the wild. Read the notes. Then look at the web that follows to see how the students identified categories for their observations and combined them in one web.

**Nita**

I saw 19 adult butterflies, 47 caterpillars (larvae), and no pupae. All the caterpillars I saw were on milkweed leaves, eating. I saw two caterpillars being eaten by birds (a blue jay and a red-winged blackbird). I saw two pairs of butterflies mating on the ground.

**Caren**

I saw 22 adults. Some were on flowers, sipping nectar. I saw 39 caterpillars eating milkweed leaves. I looked for monarch pupae but didn't see any. I saw two females laying eggs on the underside of milkweed leaves. I saw one butterfly pair mating in the air. I saw a blue jay eating an adult butterfly.

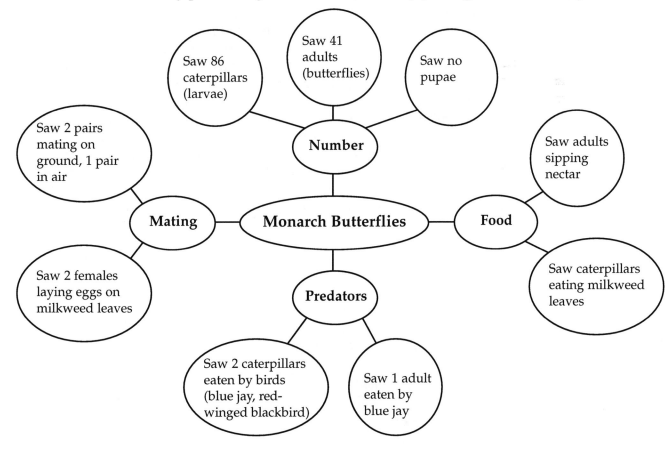

**Application**  Read the essay on animals of the upper intertidal zone. Then use the web on the following page to categorize the information in the essay. Remember, as you read, you need to identify each piece of relevant information. Next, consider the categories you will use to organize the information. Then you can add the information to the web. Finally, review your web and add any new categories or information you think of.

## Observations of Intertidal Animals

We went to the Natural Bridges Park to observe animals in the intertidal zone. Because the tide was partially in, we were able to observe only the splash zone and high intertidal zone. We looked at organisms and looked them up in a field guide. We also wrote down which of the two zones we found them in and what their behavior was like.

We started in the splash zone, the highest zone above the water, where I found some gooseneck barnacles. They were attached to the rocks with what looked like some kind of glue, but I knew it was a natural secretion that the barnacles produced so they didn't get swept away. They were closed up. They only open when they are covered with water. The secretion that attaches them to the rocks is produced on their heads, so they are head down. When they open, their legs catch plankton to eat. We also saw a few crabs in the splash zone. They were hiding under rocks, but when disturbed, they scuttled across our path. Some of them were bluish-green. Others were a rusty orange color. We also saw many limpets clinging to the rocks. They seemed completely inactive.

Then we moved down to the high intertidal zone. This area is only covered at the highest tide of the day. We found sea anemones. Some, in high tidepools, were open, and their tentacles were waving. Others were all closed up. We also saw some sponges in the tidepools and a couple of little isopods, which were either young crayfish or lobster; we couldn't tell which. Also, we saw a couple of small fish in one of the tidepools. According to our guide, they were probably young opaleyes. Elsewhere in the high intertidal zone, we saw a lot of mussels. They were also cemented to rock. They were all closed up, too, because they don't open unless covered with water. Under the shelf of a rock, we found a whole bunch of sea stars. They were multiple colors—red, purple, and orange. One of them seemed to be trying to pry open a mussel shell.

**Web** Use the information in the reading to fill in the web below. Organize it in a way that makes sense to you. Remember, some pieces of information might apply to more than one subcategory. Add or delete boxes and lines as needed.

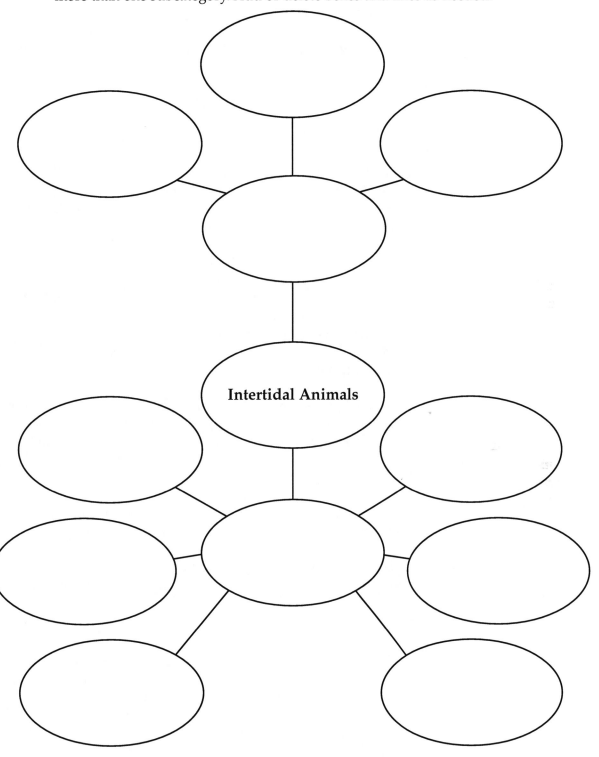

Intertidal Animals

**Tables**  Another great resource for organizing data is a table. Tables use categories to organize information. When the information you are trying to organize describes several items and attributes they have in common, a table is often a good organizer to use.

A table is simply a grid with rows and columns. Tables are useful because information stored in a table is easy to find—much easier than the same information embedded in text. Because of this, tables are helpful when you need to review material for a quiz or a test.

**Using Tables**  Tables come in all sizes and shapes. The size and shape of a table depends, of course, on what's in it. When you're reading material that you will need to recall later, the first step is to think about how to take notes on it. Should you make a table, or should you take notes in some other form? If what you're reading is a description, a narrative, or a logical argument, it may not lend itself to storage in a table.

Usually, a table has a row (the horizontal part) for each item being listed. The columns (the vertical part) provide places for aspects of the listed items—the things they have in common. The places where the rows and columns meet are called cells. In each cell we write information that fits both the topic of the row—the thing being listed—and the topic of the column—the aspect being examined. To create a table, we make rows and columns to fit the number of items and attributes.

To make a table, start by deciding what items are being listed. Next, determine the categories of information that you will show in the table. Draw a table with the appropriate number of rows and columns for the items and categories. Write the items being listed at the start of each row. Write the aspects being examined at the top of each column.

Once you have the table set up, fill in the cells with information that fits both the row and the column. Finally, give your table a descriptive title.

**Tables in Action**  Here is an article about star classification. Read the article. Then see how the information from the article has been organized in the table that follows.

## Star Classification

There are millions of stars in the sky. To study them, astronomers have found ways to classify stars. The most common classification is based on a star's spectrum, or the type of radiation it emits. Using this system, stars are grouped into nine classes. Each class has a letter as a name. In order of decreasing temperature, the classes are OBAFGKM.

The hottest stars are O-type stars. These stars look bluish. Their temperature ranges from 28,000 K to 50,000 K. Their mass is from 20 to 100 times that of the Sun. Most O stars have lifetimes of about 3 to 6

**12** © 2005 Walch Publishing

million years. An example of an O-type star is Alnitak, located in the constellation Orion.

Next come B-type stars. These stars look bluish-white. Their surface temperatures range from 10,000 K to 28,000 K. Their mass is from 3 to 20 times that of the Sun. An example of a B-type star is Regulus, found in the constellation Leo.

A-type stars look white. Their surface temperatures range from 7,500 K to 11,000 K. The mass of most A stars ranges from 1.5 to 3 times that of the Sun, but A-type supergiants may have masses up to 16 times that of the Sun. An example of an A-type star is Altair, which is in the constellation Aquila.

F-type stars look white or yellowish-white. Their surface temperatures range from 6,000 K to 7,500 K. Their mass is from 1.2 to 1.6 times that of the Sun. An example of an F-type star is Canopus, which is part of the constellation Carina, the second-brightest star in the sky.

G-type stars look yellowish. Their surface temperatures range from 5,000 K to 5,800 K. Their mass is from 0.8 to 1.1 times that of the Sun. Our Sun is a G-type star.

K-type stars look orange or reddish. Their surface temperatures range from 3,500 K to 5,000 K. Their mass is from 0.5 to 0.8 times that of the Sun. Tau-Ceti, located in the constellation Ceti, is a K-type star.

M-type stars look orange or reddish. Their surface temperatures are less than 3,500 K. Their mass is from 0.5 to 1.3 times that of the Sun. Proxima Centauri, the nearest star to the Sun, is an M-type star.

| Type | Color | Temperature Range (in kelvins) | Mass (as factor of Sun's mass) | Example |
|------|-------|-------------------|------------------|---------|
| O | bluish | 28,000–50,000 K | 20–100 | Alnitak |
| B | bluish-white | 10,000–28,000 K | 3–20 | Regulus |
| A | white | 7,500–11,000 K | 1.5–3 | Altair |
| F | white or yellowish-white | 6,000–7,500 K | 1.2–1.6 | Canopus |
| G | yellowish | 5,000–5,800 K | 0.8–1.1 | Sun |
| K | orange or reddish | 3,500–5,000 K | 0.5–0.8 | Tau-Ceti |
| M | orange or reddish | < 3,500 K | 0.5–1.3 | Proxima Centauri |

**Application** Read the essay below. As you are reading, think about possible categories for your table. Then use the information to make a table that shows the different kinds of reactions and what kinds of reactions each combination produced. Use the blank table on page 15.

## Scientific Lab Solutions

In science lab today, we looked at several chemical reactions. Some were exothermic—they gave off heat. Others were endothermic—they attracted heat from the outside. Some had no heat transfer. Those without heat transfer showed the chemical reaction by a change in color, giving off a gas, or a precipitation of material at the bottom of the container.

First, we took a flask of supersaturated sodium acetate solution and put in a sodium acetate crystal. Immediately, the whole solution started turning into white crystals. We touched the outside of the flask; it was distinctly warmer.

Next, we took barium hydroxide and ammonium nitrate in a 2:1 ratio and mixed them in a small beaker. Almost instantly, the solution got very, very cold. When we took the temperature, it was –20°C! The mixture did not change color. We did not see any precipitate.

Our next experiment was to take a solution of acetic acid and water, then add a tiny amount of phenolphthalein. The solution turned bright pink. Then we added a small amount of ammonia and water, and the color vanished. We did not notice any precipitate or any temperature change.

Our class was still not over. We combined acetic acid and sodium bicarbonate, and we observed gas rising from the solution as bubbles. This was a gas reaction; we learned that the gas was carbon dioxide. There was a slight warming of the solution, a difference of about 3°C. No precipitate appeared, and no color change occurred.

Finally, we added ten drops of aqueous lead nitrate solution to 100 mL of ammonia. We watched as the small pieces of lead precipitated out of the solution after it foamed up into a white, creamy soaplike gas. There was a slight cooling of the solution—less than a degree. In appearance, it changed from colorless to white.

**Table** Use the information in the article on page 14 to fill in the table below. Add or delete rows or columns as needed.

| Chemicals Combined | Exothermic or Endothermic? | Gave off Gas? | Changed Color? | Precipitated? |
|---|---|---|---|---|
|  |  |  |  |  |
|  |  |  |  |  |
|  |  |  |  |  |
|  |  |  |  |  |
|  |  |  |  |  |
|  |  |  |  |  |

**15**

## Problem and Solution Charts

We have looked at two types of organizers in this section, but there are many other ways to organize information. Here is another organizer you could use to brainstorm problems and solutions. Experiment with different ways to set up graphic organizers to find the ones that work best for you.

Write the problem in the horizontal box at the top. Write possible solutions in the vertical boxes. Once you have investigated the possible solutions, write the most successful one in the horizontal box at the bottom.

**Problem**

**Possible Solution 1**      **Possible Solution 2**      **Possible Solution 3**

**Best Solution**

# 3 Classifying Information

Once you have organized your information or data so that it makes sense to you, it is often time to classify it so that it makes sense to anyone else who looks at it.

All branches of science rely on classification to organize data. In fact, some branches are entirely devoted to classification. We classify everything in nature—chemical elements, physical laws, types of energy, rocks, minerals, stars, planets, plants, animals, bacteria, viruses, ancient cultures, and even human development. Classification makes it easier to

- memorize facts

- sort items by relationships

- visualize cycles in nature

- see how forces and motion work

There are thousands of ways to classify information in science. We will take an in-depth look at three relational classification schemes:

- hierarchical diagrams

- cladograms

- continuum scales

We will also look at two other ways of classifying information:

- cycles

- free-body diagrams

These graphic organizers are classification devices, although they are quite different from relational classifications. They are used to classify events over time. You have probably seen cycle diagrams before. Free-body diagrams are a way to show the forces acting on an object.

We will also take a quick look at a pyramid diagram.

Which classification scheme you use depends on what you want to emphasize in your data. Do you want to show relationships between objects in nature? Do you want to visualize forces or cycles in nature? On the following pages, you will learn how to use various classification diagrams.

## Hierarchical Diagrams

Defining the relationships between organisms or objects is an extremely important part of science. That is why scientific information is often classified. The technique used to classify depends on whether the classification is based on relationships, cycles in nature, or forces in nature. One technique often used to classify relationships in science is a hierarchical diagram.

A hierarchy is a way of showing how things relate to one another by grouping them according to some significant characteristics. Hierarchies move from the most general to the most specific. The item at the top of the hierarchy—called the *root* of the hierarchy—is the most general or most important item in the grouping. Under that come subcategories that relate to the root, but are less general or less important. At each successive level in the hierarchy, the subcategories become more specific and less inclusive. Usually, each level of the hierarchy "contains" all the levels below it.

For example, we can divide the field of science into several large subcategories: life science, physical science, and earth science. Each of these subcategories can be further divided. Subcategories for life science include biology, botany, and zoology. Subcategories for physical science include physics, chemistry, and mechanics. Subcategories for earth science include geology, meteorology, oceanography, and seismology. And then each of these subcategories can be further subdivided. For example, biology can be divided into microbiology, anatomy, and immunology. All these relationships could be shown in a hierarchical diagram.

A well-known example of a hierarchy used in science is the taxonomy used to classify organisms. The root, or broadest category, is the kingdom, followed by phylum or division, class, order, family, genus, and species. Each level is more specific, with narrower characteristics, than the level before. In a taxonomy, by the last level—the species—each subcategory has only one member.

## Using Hierarchical Diagrams

A hierarchy is usually shown as a diagram with a tree-like structure. You can use a hierarchical diagram for systems in which members have an ordered relationship. Common ordered relationships include rank, importance, sequence, and being part of a whole.

To create a hierarchical diagram, start by analyzing the information to identify the relationship. Are items related by rank, in a sequence, as parts of a whole, or through some other relationship?

Next, decide how many levels you will include. Identify the root, or broadest category, of the hierarchy. Write it in a box at the top of the diagram. Then identify the subordinate categories and the characteristics used to describe members of each category. Write the first level of subcategories in boxes below the root. Draw lines to connect them to the root. Do the same for each level you want to include, with boxes for subcategories and lines to connect them to the level above.

## Hierarchical Diagrams in Action

Here is an article about an aquarium exhibit. Read the article. Then see how the information from the article has been organized in the hierarchical diagram that follows.

### From the Shallows to the Depths

The aquarium's newest exhibit is called "From the Shallows to the Depths." It includes marine animals that live in a variety of environments.

The first part of the exhibit features animals commonly found in tidepools and shallow coastal waters. Many animals in this part of the exhibit are invertebrates—animals that have no backbone. These animals include echinoderms, such as sea urchins, sea stars, and sand dollars; and crustaceans, such as crabs and shrimp.

The second part of the exhibit is home to animals of the deeper ocean. Most animals of the deeper ocean are vertebrates, including mammals and fish. Mammals are warm-blooded vertebrates that breathe air and nurse their young—but some of them live in the water. Ocean mammals include seals and dolphins. Fish species are amazingly diverse. These cold-blooded creatures live in both fresh and salt water. Most fish have fins, gills, and scales, but not all have bones! Bony fish include angelfish, trout, and tuna. Some fish have skeletons made of cartilage instead of bone. These cartilaginous fish include sharks.

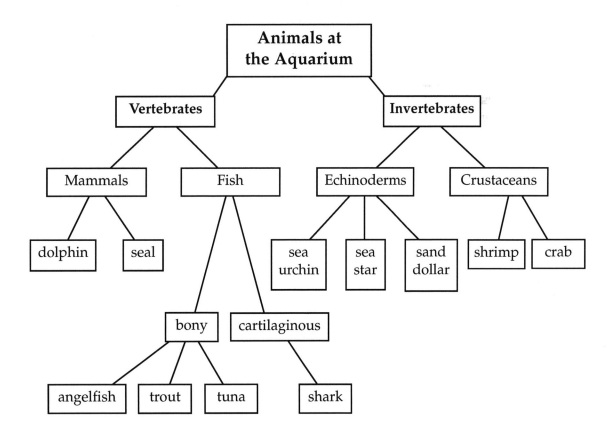

**Application** Read the essay below. Then use the hierarchical diagram on page 21 to categorize the information. As you read, think about the characteristics you want to define. How can the objects in the reading be classified together? What characteristics distinguish the objects from one another? What examples can you use to illustrate these characteristics?

# Monerans

The simplest life forms on Earth are *monerans.* Unlike more complex life forms, they do not have a nucleus. Their DNA is carried throughout the cell. The earliest monerans were anaerobic—they could not tolerate oxygen. Today, however, there are only a few forms of anaerobic monerans left. Together, they are called archaea.

One phylum of archaea is methanogens, which produce methane. They live in swamps, sewage treatment plants, and intestinal tracts. Another is the halophiles, which live in salty areas, such as the Great Salt Lake in Utah. Another phylum of archaea are the thermoacidophiles, which like hot and acidic conditions. They can live in astonishing places, such as in sulphur hot springs and around thermal vents at the bottom of the ocean.

The other large group of monerans is bacteria. One phylum is the cyanobacteria, which used to be called blue-green algae. They created most of the free oxygen on early Earth. Another phylum is the gram-positive bacteria, which are called this because they can be stained using Gram's stain. Some gram-positive bacteria cause dangerous diseases, including pneumonia. The third phylum of bacteria is the proteobacteria. They include bacteria that fix nitrogen in the soil and make plant life possible. The fourth phylum is the spirochetes, which are spiral shaped and are mostly parasitic.

Most bacteria reproduce by simple fission. They pinch off somewhere in the middle and simply divide. Because their reproductive system is so simple, bacteria are remarkably stable life forms.

Even so, they can respond quickly to changes that threaten them. Overuse of antibiotics is causing the emergence of resistant strains of some bacteria. This is a growing threat to human health.

**20** © 2005 Walch Publishing

# Hierarchical Diagram

Use the information in the reading to create a hierarchical diagram.

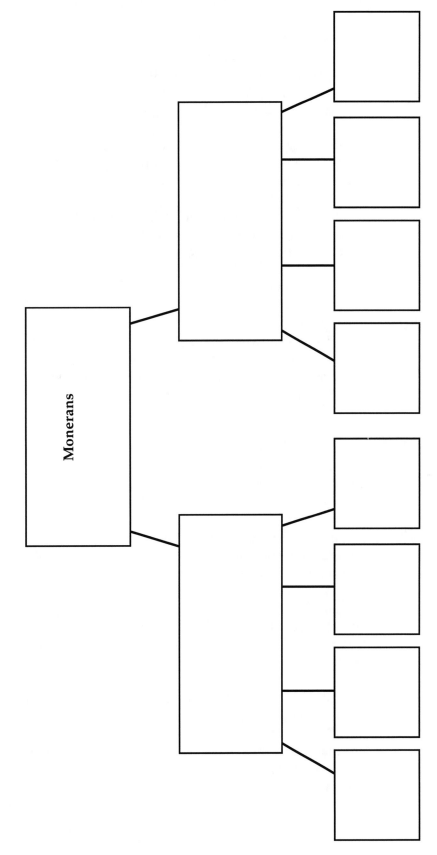

**Cladograms**  Like the hierarchical diagram, a cladogram is a way to show the relationships between organisms. Cladograms are usually only used in evolutionary biology.

The word "cladogram" comes from a Greek word meaning "branch." A cladogram is a branching diagram that looks like a series of Y's. Each fork of a Y shows where a species branched off, or diverged, from a common ancestor. Groups of organisms that share common ancestors (and therefore have similar features) are called clades.

Relationships between organisms are based on shared derived character. These are characters that all members of the group share but that did not exist in earlier organisms. Organisms can be divided into two groups based on whether or not they have a specific shared derived character. For example, early life forms on Earth needed to return to the water to lay their eggs. Then some organisms developed a leathery egg that could be laid on land. Organisms that developed these eggs shared a derived character. Organisms that still laid their eggs in the water did not. This derived character would be shown by a branch on a cladogram.

**Using Cladograms**  Cladograms are a good way to visually organize the relationships between organisms. They assume that, at different times, existing organisms divide into distinct groups. Each time an organism diverges, scientists believe that two organisms result. On a cladogram, the points where the organisms diverge are called nodes. Each node has two branches, one for each species that results from the divergence. At the end of each branch is an organism, either living or extinct.

The diagram of a cladogram starts at the bottom with the earliest common ancestor. When one organism develops a derived character that the rest of the group does not possess, a branch is drawn from the first group. The point where the branches connect is called a node.

The first node shows the earliest species divergence. At the node, one branch represents the new species, while the other represents the original species. The branching lines show the formation of two species from one common ancestor.

To make a cladogram, start by choosing a clade, or a group of organisms to be analyzed. Identify characters that are present in the different organisms. Determine whether they are primitive (inherited from earlier ancestors) or derived (recently evolved).

Next, determine whether each organism has or does not have each derived character. Group the organisms according to shared characters. (A matrix of organisms and characters is a useful way to do this.) The earliest organism is the one that shares the fewest shared derived characters with other members of the group. Write this organism at the base of the diagram, with a branch leading up from it.

**22** © 2005 Walch Publishing

Determine the first divergence, where a shared derived character develops. At this node, draw a second branch leading away from the first one. Write the name of the organism at the end of the branch. Continue this process, adding nodes and branches, until all your organisms have been placed on the cladogram.

**Cladograms in Action**

Read the article below about reptile evolution. Then look at the cladogram that follows to see how the information in the article has been shown on the diagram.

# Reptile Evolution

All reptiles, both living and extinct, descended from an unknown amphibian ancestor. Unlike amphibians, reptiles could survive away from the water. The first reptiles, known as anapsids, had a powerful, toothless jaw. Their brains were protected by a hard, bony case and an outer skull armor. The space between was filled with muscles to open and close the jaws. There were no openings in the skull behind the eyes. Many scientists believe that the anapsids were early ancestors of turtles and tortoises.

Then some organisms developed openings in the outer skull, with jaw muscles on the outside of the head. This led to the development of stronger jaws. Animals with these stronger jaws could handle larger prey. Organisms that developed two openings in the skull are known as diapsids.

Eventually, the diapsids split into two distinct groups, lepidosaurs and archosaurs. Lepidosaurs are characterized by a forked tongue. They are the ancestors of today's lizards and snakes. Archosaurs had teeth in separate sockets, rather than one long groove.

The next group to diverge were the crocodilians. They are the ancestors of crocodiles and alligators.

The next branch to break off from the archosaurs was the now-extinct pterosaur. This branch included flying and gliding reptiles.

Next came the dinosaurs. Sauropods—large, four-legged reptiles—and theropods—reptiles with strong back legs and short front legs—were part of this branch. Modern birds are descended from theropod dinosaurs.

The last branch of the reptile clade was the ornithopods. They are also now extinct. They were dinosaurs with hips shaped like a bird's hips. Even though modern birds looked more like these dinosaurs, fossil evidence suggests that they are directly descended from the two-legged theropods.

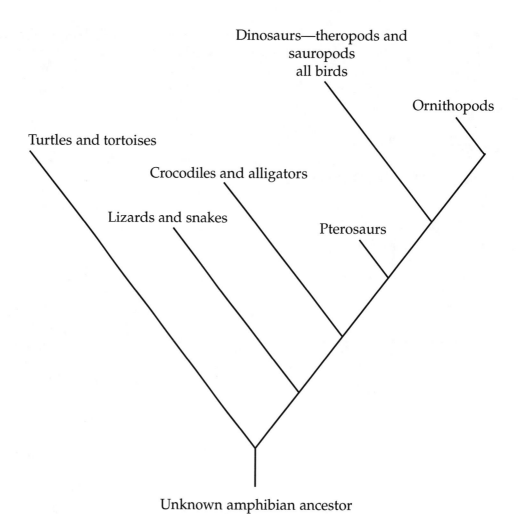

Dinosaurs—theropods and
sauropods
all birds

Ornithopods

Turtles and tortoises

Crocodiles and alligators

Lizards and snakes

Pterosaurs

Unknown amphibian ancestor

**Reptilia**

**24**

© 2005 Walch Publishing

**Application**   Read the article below about human evolution. Then use it to create your own cladogram of *Homo sapiens* on page 26.

As you read, think about the oldest common ancestor (living or extinct) for the group in the reading. Consider how each branch of descendants differs and diverges from each ancestor.

# Evolution of Humans

Human evolution is marked by incredible advancement as well as some sad dead ends. The common ancestor for all species of humans were ancient apes, mostly tree dwellers, which lived in Africa some 5 million years ago. One of them evolved into a small apelike organism called *Australopithecus afarensis*. *A. afarensis* was important because skeletal remains suggest these small creatures walked upright. This creature lived in Africa more than 4 million years ago.

After *A. afarensis*, the path diverged. One path led to *A. robustus* and its near relation, *A. boisei*. Both are now extinct. Both *A. robustus* and *A. boisei* show distinct human features, but retain the heavy bone structure of their ape ancestors. By 1 million years ago, both had died out.

The other path led to *A. africanus*, who lived about 3 million years ago. There, the path diverged again. One branch led to *Homo habilis*, who lived about 2 million years ago. That branch also comes to a dead end. *H. habilis* is best known as the first tool-making human. The other branch led to *Homo erectus* and their descendants. *H. erectus* lived around a million years ago. Their immediate descendant was the first *Homo sapiens*, called *Archaic H. sapiens*.

Once again, the path branched off. One path led to the Neanderthals, a heavy, robust human who lived in Ice Age Europe. Neanderthals looked similar to modern humans, but still retained the heavy brow ridge characteristic of the ape ancestors. Still, it is known that Neanderthals were quite advanced. They participated in burials, for instance, which suggest a belief in the afterlife, and they made extremely fine tools. It is believed that Neanderthals were wiped out, in part, by their own cousins, Cro-Magnon humans. Neanderthals died out perhaps 20,000 years ago.

The other branch led to Cro-Magnons, who lived in southern Europe and elsewhere around the world. Cro-Magnons were the ancestors of all modern human races.

Cro-Magnon fossils have been found with cave paintings and human figurines. By no later than 30,000 years ago, they were primarily responsible for the establishment of human culture. They created villages and agriculture, in addition to being the first to herd animals.

By 10,000 years ago, human written history began. The first cities were established no later than 9,000 years ago. This is the point when it is considered that modern humans began their reign. Now, we are called *Homo sapiens sapiens* to distinguish us from our early *H. sapiens* ancestors.

**Cladogram** Use the reading on page 25 to chart human evolution on this cladogram. Make sure you include all the ancestors of *Homo sapiens*. Add or delete branches as needed.

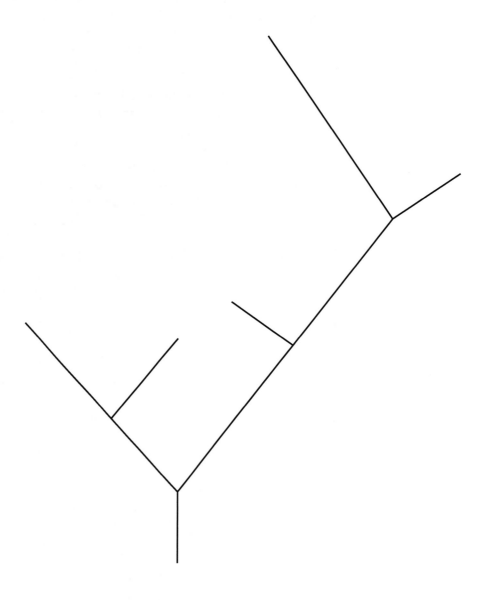

## Continuum Scales

Like the hierarchical diagram and the cladogram, the continuum scale is used to relate objects, phenomena, or events to one another.

Scales do something other organizers do not: they set limits for data. Any time you use a range of numbers to assign value, such as a 4.0 grading system in school, you are setting up a continuum scale. This grading system represents only grades of zero to 4.0. Therefore, the highest grade possible would be 4.0, and the lowest grade possible would be zero.

A continuum shows degrees of difference between two extremes. The extremes could be specific numerical values, or they could be different characteristics, such as dark and bright, tall and short, or windy and calm. For example, the Beaufort scale measures wind. A rating of 0 on the scale means that the air is still, while a rating of 12 means winds are hurricane-force.

## Using Continuum Scales

Continuum scales can be used in many ways. You can use a continuum to show how an object or several objects rank within a range. Does an animal produce more offspring or fewer than other animals? Is a ray of light closer to the blue end of the spectrum or the red? Is a piece of gypsum closer to the soft mineral end or the hard mineral end?

You can also use a continuum to show the amount of progress a person or thing has made from one end point to another, for ratings scales, or to show degrees of something—less to more, low to high, few to many. If a topic has a definite beginning and ending point but many gradations in between, a continuum can be a good way to organize the information.

Sometimes a continuum scale has internal markings. The Moh's hardness scale for minerals is one such scale. The softest mineral (talc) and the hardest mineral (diamond) are the outer boundaries of the scale. However, eight other minerals in between the hardest and softest provide guidance as to where on the scale a newly discovered mineral might fall.

Sometimes a scale has no markings, or the markings are indistinct. The electromagnetic spectrum is one such scale. There are general notions for where microwaves end and infrared radiation begins, but the waves tend to blend into one another. The scale reflects this.

Nevertheless, both the scale of hardness and the electromagnetic spectrum are continuum scales because they show how objects rank within a range.

To create a continuum scale, start by determining your limits on both ends. What is not part of the scale, and what is? Your outer limits may change as new data becomes available.

Next, decide how to measure items on your scale. This should be based on the type of information you want to show. For example, if you wanted to show the relative thicknesses of flower stamens, you would use centimeters or millimeters to measure them. If you wanted to show evolution from the first invertebrates to modern humans, you would use millions of years.

Once you have decided on the end points and the system of measurement, you are ready to draw the continuum. Your scale can be as simple as a horizontal line with end points marked, or it can be more complex, with information on both a horizontal and a vertical scale. If you need internal divisions on your scale, mark them clearly.

Once you have created the scale, just add the elements you want to track at the appropriate places on the scale.

## Continuum Scales in Action

Read the article below about acids and bases. Then look at the continuum scale that follows to see how the material has been arranged.

### Acid or Base?

An important factor in chemistry is the relative acidity or basicity of substances. This is known as the substance's pH factor. The pH scale extends from zero, the most acidic, to 14, the most basic. Neutral is 7. Distilled water has a pH factor of 7.

Testing for pH is usually done with litmus strips—strips of paper that have been treated so that they change color in either acid or basic solutions. If a solution is acidic, the strip will change to a reddish color. If the solution is a base, the strip will turn bluish. The pH of the substance can be determined by comparing the litmus strip color to a color chart.

Acids and bases are found in most households. Many cleaning fluids, such as ammonia, are bases. Ammonia has a pH of 11.5. Another common base used in the kitchen is baking soda. When mixed in a solution with water, baking soda has a pH of 8.3. There are plenty of acids in the kitchen, as well. Ordinary milk has a pH of 6.5, making it a weak acid. White vinegar has a pH of 3.0.

The strongest acid, hydrochloric acid, has a pH of 0. The strongest base, sodium hydroxide, has a pH of 14. Strong bases and strong acids are both corrosive; that is, they can burn through skin and other objects. When combined, however, acids and bases react, and move closer to a neutral pH. When working with acids and bases, always neutralize them before pouring them down the drain.

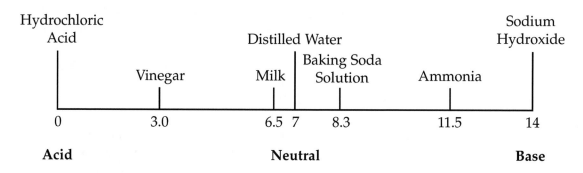

**Application** Read the essay below about the reproductive strategy continuum. Then create a continuum scale on page 30. As you read, think about what your continuum's limits will be. What measurement system will make sense for your data? Do you need internal markings, such as neutral in the pH scale?

## Reproductive Strategies

Among all life forms, there are two basic reproductive strategies, known as r-selection and K-selection. The r-selection strategy is to increase the population of the species as much as possible, without regard for whether young organisms survive to reproductive age or not. Organisms that use the r-selection strategy tend to mature early and to make millions of seeds or spores, or lay millions of eggs, during the course of their lifetimes. One or two of the organisms might survive to adulthood. The rest never germinate or hatch, or they are eaten early in their lives as part of phytoplankton or zooplankton. An example of an animal that uses the r-selection strategy is the mussel. It releases millions of eggs into the water, where they may or may not be fertilized.

The opposite strategy is the K-selection strategy, which shows a greater concern for the carrying capacity of the environment. Organisms that use this strategy reach maturity later. They focus their energy on a few offspring or spend a great deal of energy protecting their seeds. The few offspring that are produced are much more likely to survive to adulthood because the parents are more involved in assuring their survival. The young of extreme K-selection organisms tend to have a very long childhood. In animals, this childhood involves being cared for by a mother or father or both, or even by an extended family. The youngster spends its childhood learning about what is involved in adult life. An example of an extreme K-selection animal is the gorilla, which may produce one offspring every five to six years.

**Continuum Scale** Given what you know about animal life, place the following organisms where you believe they should go on a continuum from extreme r-selection to extreme K-selection.

- monarch butterfly
- zebra
- elephant
- frog
- shark

- sea turtle
- earthworm
- robin
- sea star
- human being

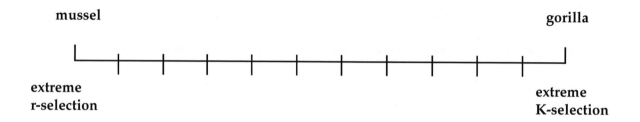

mussel

gorilla

extreme
r-selection

extreme
K-selection

**30**

© 2005 Walch Publishing

**Cycles**    In addition to relationships, scientists also examine events within a single life, or within a single phenomenon. If the events repeat, the information can be classified into a cycle.

Cycles show progression within a single phenomenon over time. They can show everything from the metamorphosis of a frog to the way rocks change over time.

**Using Cycles**    Things in nature often occur in repeating patterns. A cycle shows the progression of the pattern and how the pattern repeats. For example, a frog lays an egg, which becomes a tadpole and begins metamorphosis; intense pressures in the earth melt metamorphic rocks into magma, beginning the rock cycle again; the pattern of the Sun's magnetic field helps predict sunspot patterns.

Cycles are usually shown in a circular pattern. Occasionally, as with the sunspot cycle, they are graphed. The sunspot cycle graph shows a repeating sine wave pattern. No matter which way they are presented, they are both cycles. This is the wave pattern for the 11-year sunspot cycle:

**Sunspot Cycle**

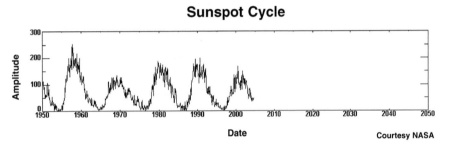

Courtesy NASA

Follow these steps to make a cycle. First, decide the most logical beginning and ending point. Determine whether you will use a circular cycle pattern or a wave. If you are using a circular cycle diagram, write down the important stages of the cycle in a circular pattern, beginning and ending with your starting point. If you are using a wave pattern, make a graph with the time frame on the horizontal scale and the unit of measurement on the vertical scale. Give your cycle a descriptive title.

**Cycles in Action** Read the article below about monarch butterflies. Then look at the cycle diagram that follows to see how the material has been arranged in a cycle.

## Life Cycle of the Monarch Butterfly

The life cycle of the monarch butterfly is one of the best-known examples of metamorphosis in science. The cycle begins when a female butterfly lays an egg on the underside of a milkweed leaf. When the larva hatches, it is called a caterpillar. The caterpillars eat only milkweed leaves. The young caterpillar gains its own weight every day. After about two weeks, the caterpillar has achieved its full size and stops eating. It hangs, head down, from a twig and sheds its outer skin, developing a chrysalis. It is now in the pupa stage. Monarch chrysalises are green; as the pupa stage continues, they slowly become more translucent. Finally, the adult butterfly within emerges. It can no longer eat leaves but can only drink nectar. The adults mate, and the female lays her egg, beginning the cycle all over again.

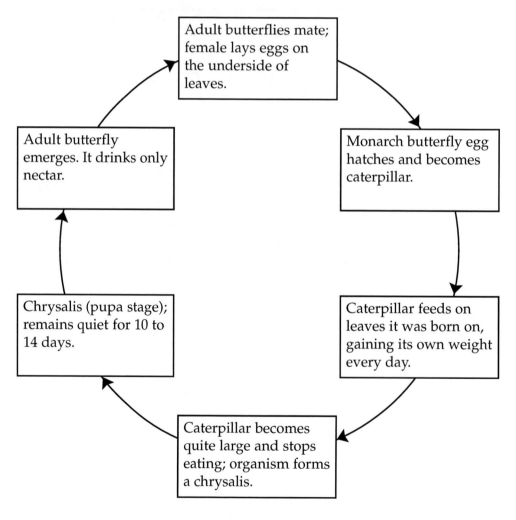

**Life Cycle of the Monarch Butterfly**

**32**

© 2005 Walch Publishing

**Application**   Read the article below about the water cycle. Then use the information to complete the cycle diagram on page 34. As you read, think about a logical place for your cycle to begin. Will you use a circular or wave cycle? What are the important stages of the cycle?

## Earth's Water Cycle

Earth's water is constantly circulating. The water that was on Earth early in its history is on Earth today, though perhaps in a different form.

The Sun's energy evaporates water from the ocean, from glaciers, and from other surface water sources, such as lakes. This causes the hydrologic cycle to begin.

Water molecules, in the form of water vapor, rise into the atmosphere. The vapor condenses into clouds. If the amount of water vapor approaches 4 percent, it begins to precipitate out of the atmosphere and fall back to Earth.

Precipitation can be in the form of rain, sleet, hail, or snow. Sleet and snow are more common when the weather is cold; rain and hail occur during warmer periods.

Oceans take up nearly three-quarters of Earth's surface, so most of the precipitation, in whatever form, falls on the ocean. If the precipitation falls on the ocean, the water cycle is complete. But not all precipitation falls on the ocean—some falls on land. This causes the cycle to take a little longer. Water may drain into streams and rivers, then slowly make its way back to the ocean.

Some water collects in lakes and ponds. If that water is used for drinking water and irrigation, it will make its way into streams and rivers as runoff. Some of the water from lakes and rivers evaporates back into the atmosphere, beginning the cycle again.

Some water sinks into the ground, where it joins a vast underground water system of aquifers and underground lakes. This is called the groundwater table, and it provides the majority of drinking water for humans around the world. As it is used for drinking and irrigation, it makes its way into streams and rivers.

Some water is absorbed from the soil by plants. The water moves through the plant, from the roots to the stems to the leaves. Some water evaporates from the leaves. This process is known as transpiration.

Still other water may get bound into ice or snow. Snow may remain at the tops of mountains until summer, when it runs off into streams and rivers. A glacier, on the other hand, can last many hundreds of years.

Eventually, all water returns to the sea, where the cycle begins anew.

**Cycle** Use the cycle diagram to represent Earth's water cycle.

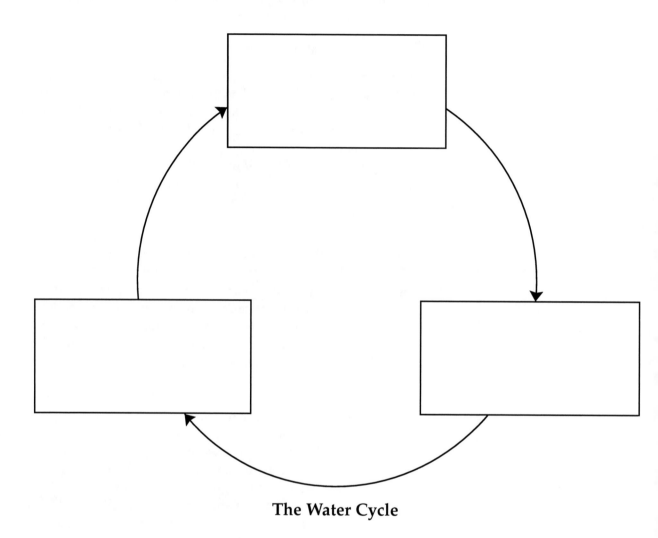

**The Water Cycle**

**34**

© 2005 Walch Publishing

## Free-body Diagrams

Free-body diagrams are simple drawings that show the magnitude and direction of all forces acting on an object in a given situation. These diagrams are used primarily in physical science.

## Using Free-body Diagrams

There are often several forces acting on a given body. If the forces are totally balanced, the body is either at rest or is in motion at a constant velocity. If the forces are out of balance, the result is acceleration.

The types of forces that can act on a body include applied force, gravitational force, normal force, and friction. An applied force is a force that is applied by a person or another object. If a person is pushing or pulling a wheelbarrow, there is an applied force acting on the wheelbarrow. On Earth, gravitational force is the force of Earth attracting objects to itself. This force is commonly called weight. Normal force is the force that keeps objects from falling into the surface they are on. Normal force is always perpendicular to the contact surface. If a book is resting on a table, the table surface is exerting a normal upward force on the book to support it. Friction force is the force exerted on an object by a surface as the object moves across the surface. Friction force is always parallel to the surface. If a box is being dragged across the floor, the floor exerts a friction force on the box.

Free-body diagrams use arrows to indicate forces acting on an object. The direction of the arrow shows the direction in which the force is acting. Each force arrow is labeled to show what kind of force it is.

Bodies are rarely at rest. They may be moving at a constant speed, or they may be accelerating or decelerating. Motion in a particular direction is a vector, and you can show the vector using vector arrows. Vectors can be added together or subtracted from one another in order to find the net motion.

To create a free-body diagram, visualize exactly what you want to show. The diagram should be simple, but may show complex force interactions. Identify all the forces acting on the object, and determine the direction each force is acting in. Then sketch simple stick figures or cartoons that can be "acted upon" by the force or the motion. Draw arrows to show each force, with the head of the arrow showing the direction of the force. Label the arrows to show what kind of force each one represents. When drawing arrows for motion, show velocity as an unbroken line ( ⟶ ) and acceleration as a broken line ( ------➤ ) where dashes show increase in velocity as acceleration increases.

## Free-body Diagrams in Action

Imagine a table with a bowl of fruit on it. The bowl is subject to a set of forces—the normal force of the table standing upright, which is holding the bowl above the ground, and gravitation, which is pulling down on the bowl. The two forces are equal but opposite, and the bowl neither falls to the ground nor flies into space. The net force is zero.

Now imagine that the forces are altered slightly—the table is lifted at one end, disrupting the balance between the downward force of gravity and the upward force of the table. The bowl begins to accelerate toward the opposite end of the table and onto the floor, with predictably disastrous results. The gravitational force became stronger than the force of the table acting upon the bowl, because the direction of the normal force changed relative to the bowl.

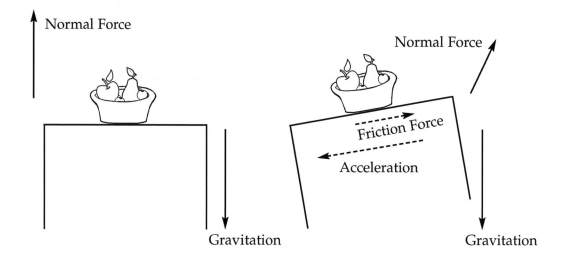

**Application**  Use the story below to complete the series of free-body diagrams on page 38. As you read, think about how you will illustrate the forces and the motion being described.

## Tug-of-War Tournament

In summer camp, we had a boys versus girls tug-of-war tournament. We held five matches on five days. We were evenly matched—six girls and six boys, and we were all going into seventh grade in the fall.

On the first day, the girls were fresher and stronger. We had just returned from a three-hour canoe trip, so the girls' combined force was greater than ours. With one mighty pull, they pulled us all over the line drawn in the grass in the center.

On the second day, we won, but it was not an easy contest. We pulled for more than five minutes, and only barely dragged one girl over the line.

On the third day, we managed to play right after the girls got back from a nature hike that took them all morning. We won that game easily by pulling all the girls over the line.

I don't even want to talk about the fourth day. We held our match first thing in the morning, and I think Pete let go to rub his eyes or something. The girls yanked us all across the line. Not only that, but it had rained the night before, and we fell in a mud puddle. One of the counselors laughed at us.

So, it all came down to the fifth day. We decided to hold the match just before we got on the bus to go home. The girls all wore team T-shirts. We, of course, hadn't thought about anything like that.

The girls stared at us, trying to psych us out. We thought Jerry might have been getting a little nervous, so we put him at the back of the rope. Then the contest started.

We pulled as hard as we could for a few seconds, hoping to throw one of the girls off balance. They, however, had dug in their heels and just let us pull until we got tired.

Then, as one, they pulled back. We sailed over the line, and the girls won the tournament.

We were disappointed, but in general, I think we did pretty well. And there's always next summer.

**Free-body Diagram**  Below are five lines, one for each day of the tournament.  Use arrows, cartoons, and whatever else you need to show the forces and the motion of the rope for each day.

**First Day**

**Second Day**

**Third Day**

**Fourth Day**

**Fifth Day**

**Pyramid Chart**   Below is a blank pyramid chart. These charts are used when a topic can be divided into subgroups, some of which contain more items than others. For instance, a trophic level chart is usually shown as a pyramid, with producers such as plants at the bottom, first-order consumers (herbivores) at the next level up, second-order consumers (carnivores) above them, and third-order consumers (carnivores) at the top.

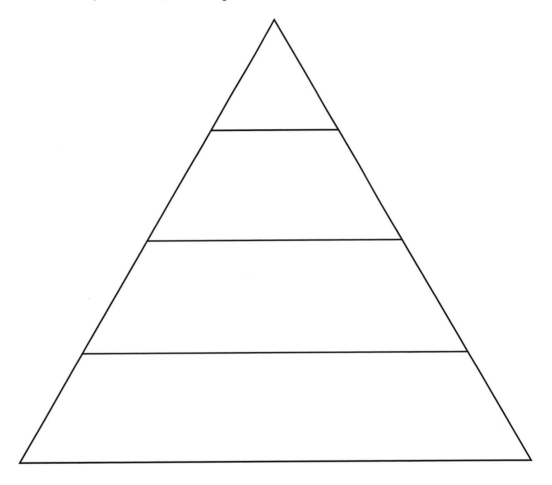

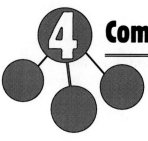

# Comparing and Contrasting

Has your teacher ever asked you to compare and contrast the characteristics of reptiles and amphibians? Or dolphins and whales, or Jupiter and Earth? Whenever you are asked to compare and contrast, you are being asked to show how things are similar and how they are different.

In the course of your science studies, you will be using data that you have collected, organized, and classified. An important way to use this data is to find out where there are similarities and differences between your data and other data, or between data taken today and data taken yesterday or last month.

A good way to show similarities and differences is with a graphic organizer. Several different graphic organizers can help you show how things and events are similar and how they are different.

Compare and contrast graphic organizers are used to

- show where two sets of data overlap

- show where there are no common data points

We will be using a graphic organizer called a comparison matrix, and we will take a quick look at a Venn diagram, which lists characteristics of two sets of data (or two organisms, phenomena, and so forth). Where the circles overlap, the data overlaps.

## Comparison Matrixes

Like a Venn diagram, a comparison matrix is used to show similarities and differences between two or more things. Unlike the Venn diagram, which can also be used as a brainstorming device, the comparison matrix lists specific attributes.

A comparison matrix is typically used when:

- Brainstorming is complete, and there are several known attributes available to compare.

- Attributes are known in advance.

- Not all attribute information is required for your purposes.

## Using Comparison Matrixes

To create the matrix, start by deciding what items you want to compare. Next, identify the attributes you want to compare. Then create a grid, similar to a table, with a column for each attribute and a row for each item. Across the top, as column heads, list the characteristics you want to compare. Along the left side, as row heads, list the objects you are comparing. Each space where a row and a column meet is called a cell.

To fill in a comparison matrix, look at each object you are comparing. For each attribute, ask whether or not the object shares that attribute. If it does, place a check mark (✔) in the cell where the row for the object and the column for the attribute meet. If the object does not have that attribute, put an **X** in the cell.

Read the article below about chordates. Then look at the comparison matrix that follows to see how the material has been arranged in the matrix.

# Chordates

One important phylum in the Kingdom Animalia is the phylum *Chordata*. We are probably most familiar with this phylum because it contains all the classes of animals with backbones—fish, amphibians, reptiles, birds, and mammals. It also contains a class of animals that possess a nerve cord early in their development but later lose it. These are the sea squirts.

All chordates possess a central nervous system and have a nerve cord at some point in their development. Another important feature of chordates is that they all possess the ability to process oxygen directly from the air or water. They might have lungs, or gills, or both. Chordates are all heterotrophs. They eat other organisms to stay alive. Some eat plants. Some eat other animals. Some eat both.

Some chordates live in the water, such as sea squirts and fish. Others live on land, such as most reptiles, mammals, and birds. Others are at home both on land and in the water, such as amphibians and some reptiles.

Some chordates lay eggs, either hard-shelled or soft and gelatinous. Some chordates give live birth. Some chordates have scales. Others have smooth skin. A few have feathers. Still others have hair.

Most chordates have a spinal column protecting their nerve cord. These are called vertebrates. Only sea squirts have an unprotected nerve cord. They are invertebrates.

Some chordates are endothermic. They regulate their body temperature through their metabolism. Others are ectothermic. They regulate body temperature through contact with their environment.

|  | Invertebrates | Fish | Amphibians | Reptiles | Mammals | Birds |
|---|---|---|---|---|---|---|
| Central nervous system | ✔ | ✔ | ✔ | ✔ | ✔ | ✔ |
| Lives on land | ✗ | ✗ | ✔ | ✔ | ✔ | ✔ |
| Lays hard-shelled eggs | ✗ | ✗ | ✗ | ✔ | ✔ | ✗ |
| Gives live birth | ✗ | ✗ | ✗ | ✗ | ✔ | ✔ |
| Has a nerve cord at some stage | ✔ | ✔ | ✔ | ✔ | ✔ | ✔ |
| Has feathers | ✗ | ✗ | ✗ | ✗ | ✗ | ✔ |
| Is heterotrophic | ✔ | ✔ | ✔ | ✔ | ✔ | ✗ |
| Is ectothermic | ✔ | ✔ | ✔ | ✔ | ✗ | ✗ |
| Has hair | ✗ | ✗ | ✗ | ✗ | ✔ | ✗ |
| Has scales | ✗ | ✔ | ✗ | ✔ | ✗ | ✗ |

**Application**  Read the essay about crystals, then create your own comparison matrix. As you read, think about the objects you want to compare and the characteristics of those objects you will be comparing.

## Types of Crystal Structure

In geology, a crystal structure is the special arrangement of atoms within a crystal. The atoms are arranged in a particular way, which is repeated along a ladderlike structure called a lattice. This arrangement, called the unit cell, can be repeated many, many times. The unit cell defines the type of crystal structure a mineral has.

The simplest type of crystal is a cube. If you look at sodium chloride crystals under a microscope, you can see them shaped in a simple, symmetrical cube. If you were to heat up water and saturate the water with salt, you would eventually build larger cubic structures than the ones in your salt shaker. The cubes might join together in such a way that there would be more than six faces, but the symmetry would be the same. There are several types of lattices that can form in a cubic system.

Another type is hexagonal. This type of crystal forms a hexagonal shape. Graphite, the material used in pencils, is a hexagonal type of crystal.

The next type of crystal is called tetragonal. These form in rectangular prisms. Many semiprecious stones fall within this category, such as cat's eye quartz.

Another type of crystal formation is the trigonal, or rhombohedral, form. These crystals form as slightly off-centered rectangular prisms, or rhomboids. Minerals in this group include most types of quartz.

The orthorhombic system is the next group. It makes small pyramids within its structure. Sulfur is an important orthorhombic crystal.

Monoclinic crystals form in rectangular prisms, where one side is either larger or smaller than all the others. The mineral gypsum takes this form.

The last and least symmetrical type of crystal formation is the triclinic group. None of its lengths are equal. One mineral with this shape is turquoise, often used in jewelry.

© 2005 Walch Publishing

**Comparison Matrix** Use the comparison matrix below to record the information about crystal structures. Across the top, list the characteristics you are comparing. Down the left side, list the objects you are comparing. Put a check mark (✔) or an ✗ in each cell to indicate whether or not an object possesses each characteristic. Add or delete columns and rows as needed.

**45** Content-Area Graphic Organizers: Science

**Venn Diagrams** The comparison matrix is used to compare specific characteristics of many objects. However, if you have just two or three characteristics and several objects, or vice versa, a Venn diagram is a great tool for comparing and contrasting.

Below are two samples of Venn diagrams. In the first, the two circles can represent two objects you want to compare, with the overlapping part representing what they have in common and the nonoverlapping parts listing which characteristics are specific to each object.

**Using Venn Diagrams** The second Venn diagram is a great way to list objects with varying characteristics. In the nonoverlapping space, you can list objects with just the characteristics of A, B, or C. In the overlapping areas, list the objects with those combinations of characteristics.

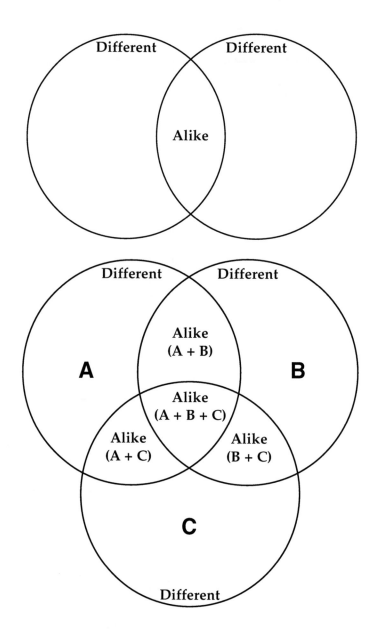

**Venn Diagram** Label each circle. Write things that are unique to each topic in the open area of each circle. Write things both topics have in common in the intersecting area.

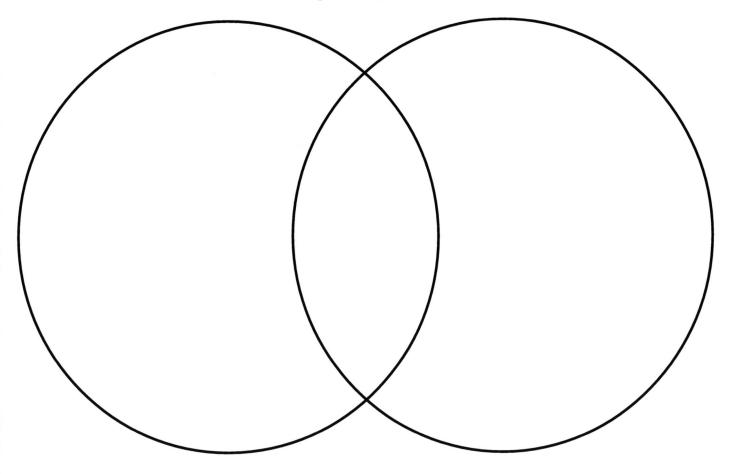

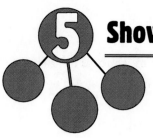

# Showing Cause and Effect

A cause is an event that makes something happen. An effect is the thing that happens. Like comparing and contrasting, showing cause and effect is a way to examine data that has been collected, organized, and classified. However, cause-and-effect diagrams show how your data has changed over time, what caused the changes, and what changes your information caused.

Use cause-and-effect graphic organizers when you want to illustrate how one phenomenon or event causes one or more effects and how several phenomena cause one effect.

We will take an in-depth look at two graphic organizers—the multiple-effect map, which shows multiple effects from one event, and the fishbone map, which shows how multiple causes work together to create one effect.

We will also briefly examine a jellyfish map, which shows how a single event leads in a number of different directions.

## Multiple-Effect Map

You are probably familiar with a basic cause-and-effect chart, in which a simple cause creates a simple effect. However, one cause can set off multiple chains of events. To show this type of relationship, we use a multiple-effect map.

This is a map that shows causes and effects over time, including chain-reaction events. It is not a snapshot of immediate results. You can use a multiple-effect map when

- you want to track chain reactions over a long period of time

- you want to see how one event triggers multiple reactions

## Using Multiple-Effect Maps

This is a particularly good graphic organizer to use when studying the history of science, the relationships between society and science (or technology), and tracking invention chains. It can be used in nearly every branch of science and is a very important tool in chemistry, sociology, anthropology, and psychology.

To create the map, you must first choose some defining event, or cause. This event must be able to be traced to several effects, some of which may trigger other effects.

The steps in creating the organizer are very simple. First, define your primary cause and list it at the top. Next, using arrows leading from your main cause, list the effects stemming from the cause. If any of these primary effects has one or more effects, list those in the same way, until you reach a logical conclusion, the end of your data, or the limit of your assignment.

Read the article below about the Toledo Library. Then look at the multiple-effect map that follows to see how the material has been arranged in the map.

## The Toledo Library

The Toledo Library was spared the worst of the Dark Ages in Europe because it remained under the control of the Moors, the Arabic people of Northern Africa. Toledo was a multicultural city. In addition to its Arab population, it had a large population of Jewish scholars. Unlike the library at Alexandria, Toledo's library was not burned in the early Christian era.

When Christian crusaders finally conquered Spain in the twelfth century, the new rulers protected both the Arab and Jewish populations and kept the library intact, at least while the works were being translated into Latin. Within the library, the crusaders found works of Aristotle, an ancient natural philosopher. His works eventually became the basis for the science curriculum at the early medieval universities at Bologna, Paris, and Oxford. Scientists studying his work during the Enlightenment tried to prove his assertions and used them to create the scientific method.

The crusaders also found scrolls that told what the ancients and the Arab peoples knew about mathematics. This information introduced Europeans to Indo-Arabic numerals, including zero as a placeholder—the number system we still use today. The knowledge about geometry found in the library was the underpinning for medieval architecture. It allowed cathedrals to soar to new heights and enabled the building of large palaces. As soon as European mathematicians mastered the algebra found in the scrolls, they went on to create more advanced forms of mathematics based on algebra, including Newton's calculus.

© 2005 Walch Publishing

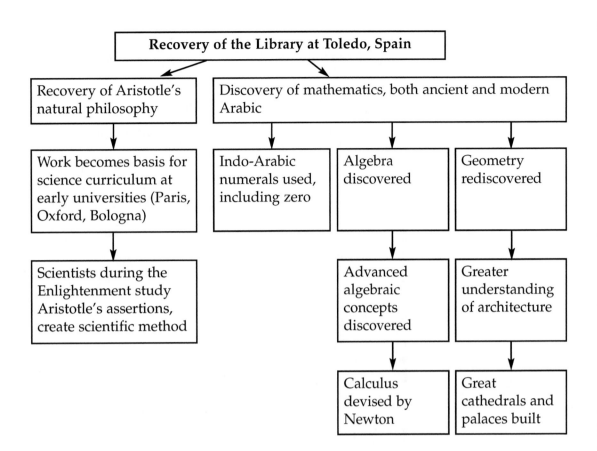

Recovery of the Library at Toledo, Spain

Recovery of Aristotle's natural philosophy

Work becomes basis for science curriculum at early universities (Paris, Oxford, Bologna)

Scientists during the Enlightenment study Aristotle's assertions, create scientific method

Discovery of mathematics, both ancient and modern Arabic

Indo-Arabic numerals used, including zero

Algebra discovered

Advanced algebraic concepts discovered

Calculus devised by Newton

Geometry rediscovered

Greater understanding of architecture

Great cathedrals and palaces built

Read the article about deserts. Then use it to fill in the multiple-effect map on page 53. As you read, think about a primary cause and the effects to which it leads.

# Spreading of the Sahara

Deserts are believed to have evolved in places that once had a lot of water. The Sahara, for instance, was covered with ice during one of the Earth's early ice ages. When the ice finally retreated, valleys and rivers were left behind. The Nile river valley shows what the landscape must have been like 10,000 years ago, even in the most desolate areas of the Sahara today.

During its fertile period, the Sahara was home to a wide variety of wildlife and even human beings. However, the air pressure over the Sahara is very high, and that causes a dry and hot climate. Eventually, the once-fertile river valleys, lakes, and inland seas began to dry out. As the desert began to advance, the region's populations retreated south.

As they went, the grazing animals, both domesticated and wild, overgrazed the marginal lands south of the Sahara. Humans contributed to the advancing desertification of the region with poor farming practices. In addition, widespread droughts were recorded in the area in the early 1970s. Although the droughts in the 1970s were the largest people had seen in recorded history, droughts are a fact of life in sub-Saharan Africa.

Another natural issue in the desert is erosion. Erosion in the desert is caused by high winds. In rocky deserts, like those in the American Southwest, wind erosion creates huge standing stones, called pediments. If the slope is steep and it flattens out on top; this pediment is often called a mesa.

Unlike rocky deserts, the sand seas of the Sahara are constantly reshaped by the hot, dry winds. The great sand dunes of the Sahara are created as wind swirls around an obstacle—usually a stone. Sand drifts in on the wind and builds the dune.

Desertification increases as the sands advance, blown by the winds from farther north. There have been many cases where small farmers, working on marginal land, emerged from houses after a windstorm to find entire crops covered with sand.

When this occurs, many small farmers give up, pack up their remaining animals and tools, and move farther south, again bringing with them the poor agricultural practices and overgrazing animals that partly caused the land they left to be reclaimed by the desert.

Stopping the advancement of the Sahara, or any desert, is a difficult task. Better farming practices must be taught and followed. Animals must be rotated to new grazing lands, allowing older ones to grow back. Crops must also be rotated, and moisture-loving crops must be abandoned altogether. Egyptian cotton, for instance, has a high water requirement. The water for its irrigation is taken from the Nile river at such a high rate that the headwaters of the Nile cannot replenish it fast enough.

**Multiple-Effect Map** Use the blank multiple-effect map below to chart the data on deserts. Write your primary cause at the top and the primary effects in the row below that. Include all the successive effects mentioned in the reading. Add or delete boxes as needed.

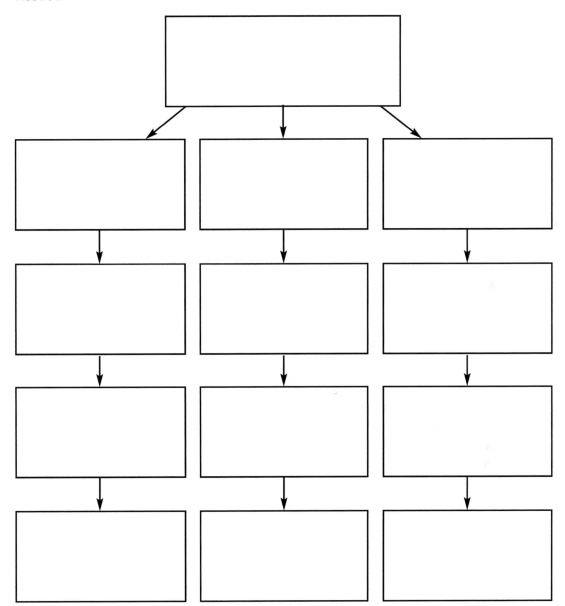

## Fishbone Maps

In the last section, you learned how to illustrate a single cause that yields multiple effects. Another type of cause and effect map shows how multiple causes create a single effect. To show this kind of relationship, you can use a fishbone map.

Like the multiple-effect map, the fishbone map shows a chain of events that occurs over time. You can use the fishbone map when

- you want to show the contributing causes to a particular effect

- you want to show how various causes lead to a specific event

## Using Fishbone Maps

Like the multiple-effect map, the fishbone map is particularly good to show events in scientific history, the relationship between science and society, and events in the fields of anthropology, ecology, sociology, and psychology. It is also useful in physics, especially in the reconstruction of physical events. For instance, this type of graphic organizer is often used to reconstruct car accidents.

The goal in a fishbone map is to show when various causes occur in relation to one another and what the final outcome of the various causes is. It is called a fishbone map because of its shape.

To create the map, you need to have some event's history you want to trace. In other words, the effect is already known. First, list the effect on one side, and draw a horizontal line from the effect across the page. Your causes will branch off either side of that line, in sequential order. The primary causes should be listed at the opposite end of the line from the effect. The later causes are listed closer and closer to the effect. Your mission is to uncover all the causal relationships to the effect and to show the order of their occurrence.

Read the article below about the death of the dinosaurs. Then look at the fishbone map that follows to see how the material has been arranged in the map.

## Death of the Dinosaurs

Scientists know, from fossil evidence, that dinosaurs died out very suddenly about 65 million years ago. During the period leading up to the eventual end of the dinosaurs, most of Earth's climate was becoming cooler and drier. New species of plants, especially grasses, were replacing the food supply the herbivorous dinosaurs ate. Grasses contain silica, which is used to make glass. Most of the herbivores could not digest the grasses, and if they tried, they damaged their teeth due to the abrasive nature of the grass.

The fossil record shows a couple of interesting things about 65 million years ago. The first is a layer of iridium, an element which does not occur naturally on Earth. Iridium is associated with asteroid strikes, or comet impacts. The next thing the record shows is a layer of fine dust and ash, which is also common in fossil records when a large volcano has erupted locally. The layer of ash and dust is quite thick and can be found worldwide, so it cannot be associated with a single volcano.

A comet, or, more likely, asteroid strike, would have thrown an amazing amount of dirt from Earth's surface into the air. It would have also touched off massive forest fires. These would have blocked the Sun's light from reaching Earth for many months. Plants died out, then herbivores starved to death.

As the herbivores died, those animals dependent upon them for food would have also begun to starve. Shortly after the layer of iridium and ash, no more dinosaur fossils were found in the record.

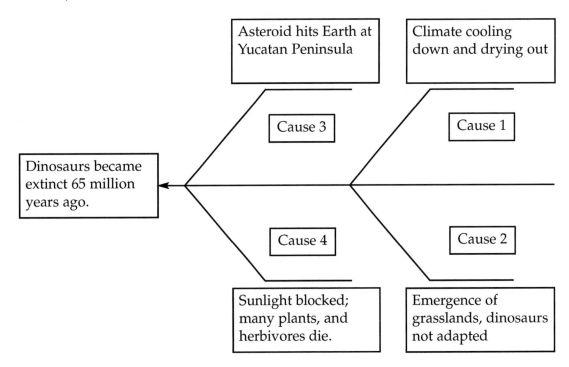

**Application**  Use this reading on early human migration to fill in the fishbone map on page 57. As you read, think about the primary effect and the causes that led to it. Also keep in mind the sequence of those causes.

# Early Human Migrations

No one knows for sure when the first human beings left Africa to begin their restless journey around the world, but some of the causes for the emigration are known. Until 100,000 years ago, all known *Homo sapiens* were located in the region of Africa in which they had evolved. The Great Rift Valley provided a nursery for the young species. Within the Valley, there were easy game animals, nearly year-round gathering opportunities for fruits and berries, and natural rock shelters that provided safe homes.

However, the species was expanding, and too many individuals were trying to survive on too few resources. Water was likely a serious problem. As elsewhere in Africa, even today, rain falls only seasonally. Animals are forced to migrate when water supplies dwindle. So, too, did the human hunters that followed them. With more humans competing for water sources, however, the water supplies wouldn't have lasted quite as long, and the young of the species would have been ill-equipped to travel.

Also, humans had managed to reduce the animal population. Undoubtedly, humans contributed to the extinction of several species of large game animals.

At the time of the first migrations, the Mediterranean Sea, which currently separates Africa from Europe, was dry. No doubt some humans migrated, following herds across the soon-to-be inundated Sea, and they became trapped on the other side after the inundation.

Southern Europe offered things that Africa did not. Regular rainfall, resulting in a regular water supply, and running rivers were very likely to have been important draws. Fewer large predators on the European continent meant that young humans were more likely to reach reproductive age themselves.

Of course, there were drawbacks. One large drawback was the change of seasons in Europe. In Africa, the temperature was mild all year long, but as the ice age advanced, living in Europe meant learning to store food for the winter, making warm clothing, and building warm shelters as protection from the ice and snow.

Eventually, European humans moved elsewhere—Asia, Oceania, and finally the Americas. Perhaps the most important reason humans migrate is the restless desire for change. Even today, humans are struggling to move to places considered inhospitable to human life—the Antarctic, the tops of mountains, and outer space.

**Fishbone Map** Use the information from the reading to fill in this fishbone map. Write the primary effect in the box on the left. Write your primary cause on the top right branch, the next cause on the bottom right branch, and so forth. Add or delete branches as needed.

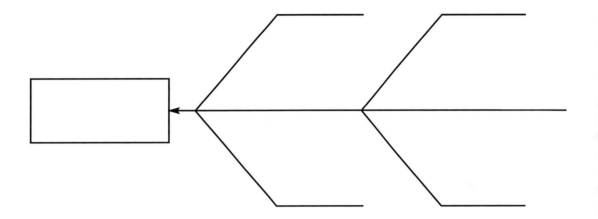

**Jellyfish Map** This type of map is often used to show how one cause or event can result in multiple chains of effects. It is a variation of the multiple-effect map. Write the cause in the horizontal box at the top. Write the effects in the vertical boxes. Add or delete boxes as needed.

# 6 Laboratories

Another important part of scientific organization is organizing laboratory reports. Lab reports are all based on the scientific method, a means of testing assumptions about the natural world.

Questions often arise during classification or creating comparison or cause-and-effect charts. When they do, the next step is to make a hypothesis consistent with the observations you have made, and then devise a way to test the hypothesis.

Creating laboratory reports makes it easier to

- develop a testable hypothesis

- create a procedure that others can also follow

- detail a list of materials

- show experimental data

- show conclusions

- show uncertainties

Unlike classification schemes, there truly is only one laboratory procedure that is accepted throughout the scientific community. In the next few pages, we will explain what the scientific method is and why it is so important to science.

However, uncertainties may turn up during the course of an experiment that may cast some doubt on the results of the experiment. Every experiment has some uncertainty in it, ranging from the well-known (limitations of equipment) to less well-known (human bias). Another chart we will look at is the uncertainty chart.

We will also look briefly at a reaction chart, useful in chemistry and physics.

## The Scientific Method

The scientific method is a process through which scientists establish verifiable information.

Today, we accept that science follows rules and processes, making it accurate, dependable, and as free from individual bias as possible. This was not always the case. Before René Descartes developed the scientific method in 1637, science was not a discipline separate from philosophy. The natural philosophers used logic, and often what *seemed* right was considered fact. Although experimentation had been done in ancient times, it had largely been abandoned during the Middle Ages.

## Using the Scientific Method

An example of prescientific-method reasoning is the notion that nonliving material can give rise to living things. People noted mold and maggots on decaying meat and plant material and concluded that mold and maggots spontaneously appeared. Since everyone believed this, there was no attempt to preserve foods better—there was no perceived reason to do so.

After the scientific method was written, in 1688, an Italian naturalist named Francisco Redi performed an experiment. He left meat in one container, open to the air, and left meat in another container, sealed tightly. Only in the open container did insect life and molds flourish.

The scientific method occurs in six basic steps:

- From your observation, formulate a question you want to explore.

- Develop a statement, called a hypothesis, that can be tested. The hypothesis is based on your observation, and it is your best guess as to what the result of an experiment will show.

- Write a list of materials you need to perform this experiment.

- Write a detailed procedure for this experiment.

- Record data accurately throughout the experiment.

- Note your conclusion. Is it the same as your original hypothesis?

**60**

© 2005 Walch Publishing

**The Scientific Method in Action**

Imagine that you are conducting an experiment about the water needs of a particular plant. You believe it needs 30 cc of water per day to be healthy. That is your hypothesis. You design an experiment involving two healthy plants of the same type, and you begin the experiment. You note the growth of the plant every week. At the end of four weeks, you make your conclusion. Was your hypothesis correct or incorrect? Take a look at the scientific method for this experiment. This graphic organizer can be used as the basis for a formal lab report.

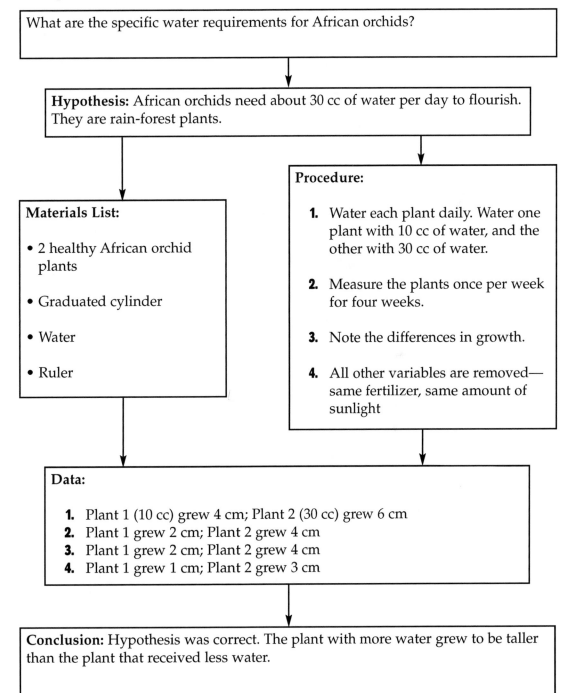

What are the specific water requirements for African orchids?

**Hypothesis:** African orchids need about 30 cc of water per day to flourish. They are rain-forest plants.

**Materials List:**

• 2 healthy African orchid plants

• Graduated cylinder

• Water

• Ruler

**Procedure:**

1. Water each plant daily. Water one plant with 10 cc of water, and the other with 30 cc of water.

2. Measure the plants once per week for four weeks.

3. Note the differences in growth.

4. All other variables are removed— same fertilizer, same amount of sunlight

**Data:**

1. Plant 1 (10 cc) grew 4 cm; Plant 2 (30 cc) grew 6 cm
2. Plant 1 grew 2 cm; Plant 2 grew 4 cm
3. Plant 1 grew 2 cm; Plant 2 grew 4 cm
4. Plant 1 grew 1 cm; Plant 2 grew 3 cm

**Conclusion:** Hypothesis was correct. The plant with more water grew to be taller than the plant that received less water.

**Application**  Read the essay below and organize the information into the scientific method format on page 63. As you read, think about what the hypothesis is, what materials were required, how the procedure was established, what data was produced, and what the results were.

# Cyanobacteria and Oxygen Lab

We conducted an experiment to see whether oxygen was essential to the life of cyanobacteria.

I did not believe it was.

We took some cyanobacteria we had collected at the pond and put some in water that we had let stand for many days until all the air bubbles were gone. Additionally, we kept a goldfish in the beaker for a few days, reasoning that the fish would use up some of the oxygen. We removed the fish and put it back in its tank. We covered the beaker, and used a straw to pull out any air above the waterline.

Then, we put some other cyanobacteria into a second beaker with fresh water in it. We did not remove the surface air.

We checked the growth of the cyanobacteria on a daily basis and decided to check the length of the strand by using a ruler marked with millimeters. On the first day, both strands were the same length—13 mm.

On the second day, the length of the strands had both increased to about 15.5 mm. They appeared to be the same length.

We checked the length each day for five days. On the third day, the length had increased to 17 mm in the beaker with no oxygen, and 17.1 mm in the beaker with oxygen.

On the fourth day, the cyanobacteria in the beaker with no oxygen was a little longer than the one with oxygen. The no-oxygen cyanobacteria was 17.5 mm, and the oxygen was 17.4 mm.

On the fifth day, both strands were 18.1 mm.

We returned the cyanobacteria strands to the pond water exhibit. It was clear that oxygen was not a necessary component for cyanobacterial life.

## The Scientific Method

Use the sample experiment to complete the organizer below.

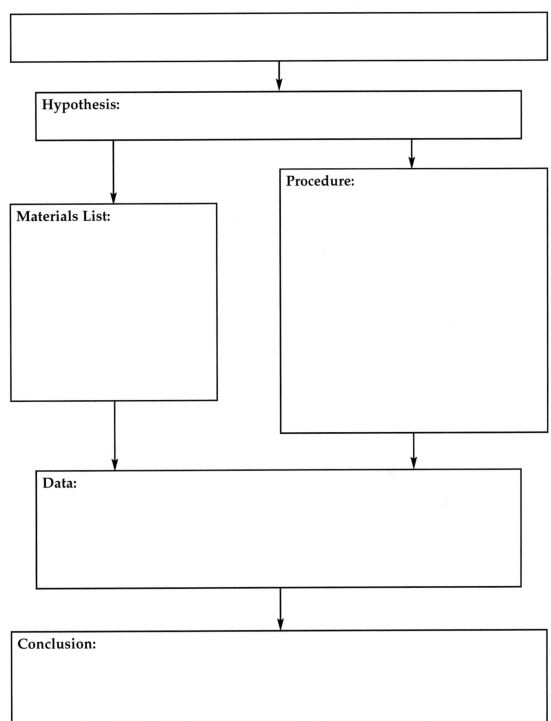

## Uncertainty Charts

There are many reasons an experiment may not go smoothly. The most important thing to keep in mind when conducting an experiment is the uncertainty that might exist.

Uncertainty occurs in all experiments, although the goal is to limit uncertainties wherever possible. Uncertainties are questions about the measurement of data in an experiment, or known or unknown factors that might affect how data is collected. Why do uncertainties occur? Here are a few examples of common uncertainties:

- Inaccurate measuring device—for instance, a scale that always seems to measure a few grams more massive than an object really is

- A measuring device that is not precise enough for the job—for instance, measuring changes in millimeters with a centimeter-ruled meterstick

- Human error

- Uncontrolled changes in temperature, pressure, or other physical conditions that may or may not alter your experiment

The two types of uncertainties are human factor and physical factor. Human factor uncertainties are introduced by humans running or taking part in the experiment. Physical factor uncertainties are introduced by anything other than human factors. You should be able to demonstrate, in a complete lab report, any uncertainty you are aware of and the possibility of uncertainties that you are not aware of.

## Using Uncertainty Charts

These are the steps to completing an uncertainty chart:

- Separate human factor uncertainties from physical factor uncertainties

- Mathematically determine how much of a problem the uncertainty may be. (Human error uncertainty is generally considered to be about 0.05%; the rest you can compute based on the range of possible error.) Add the human factor uncertainty to the physical factor uncertainty

- Multiply your assurance in your result (generally 100% in the types of lab exercises you will be doing) by your possible uncertainty. The result is the margin for error. Subtract this result from 100 to get your experiment assurance

## Uncertainty Charts in Action

In order to calculate a physical uncertainty, take the known problem (for instance, you measured a strand of cyanobacteria in mm, but you only had a cm ruler) and figure out the largest range of possible error. The largest range is 0.10, since one centimeter is ten times one millimeter. Add that to your human error uncertainty, and you have an uncertainty of 0.15. When you multiply, you find that the range of uncertainty is about 15%, which is pretty high. Next time, you might want to get your physical uncertainty down by using more precise equipment.

See how the chart for this situation is created below.

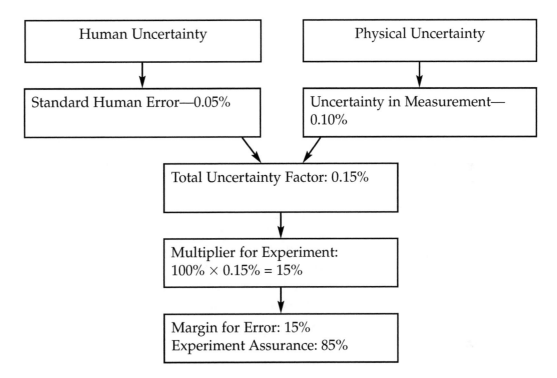

**Application** Read the description of the experiment, and organize the information as an uncertainty chart on page 67. As you read, think about what the human and physical uncertainties are.

# Radio-Controlled Car Experiment

We designed an experiment to see the velocity of a particular radio-controlled car.

Our goal in the experiment was to learn what the velocity per second was. We used a radio-controlled car operated by one of the students, a stopwatch operated by a second student, and a person who marked the distances with small pebbles.

After we measured the distance and the time, we divided the change in distance over the change in time to find our velocity, $v$. Velocity was measured in cm/s.

The car ran alongside a meterstick with centimeter markings. Since we were measuring in centimeters, that was probably fine.

We had two humans who might have introduced uncertainties into the experiment. The first was the stopwatch reader, who called "time" every two seconds. She might have called time early or late. The other was a person kneeling on the ground, who dropped small pebbles next to the meterstick every time the reader called time. He might have dropped the pebbles early or late, too.

Our little car seemed to be in good shape and running in a straight line. We didn't see any real problems with the car.

One other problem might have been that the meterstick looked like it was bowed a little bit. It probably had a tiny deviation, possibly 0.02%.

We are pretty sure of our results, which show that our radio-controlled car can move with a velocity of 40 cm/s, but we still need to factor in our human and physical uncertainties.

**Uncertainty Chart** Use the reading to complete the uncertainty chart below.

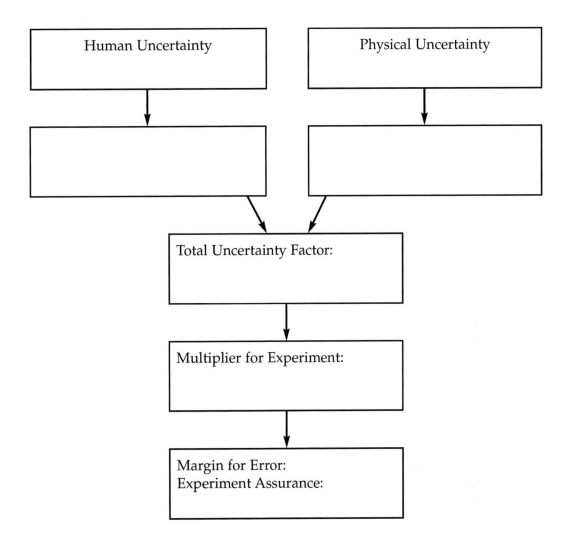

**Reaction Chart** This is a reaction chart, which may prove useful in physics and chemistry lab classes. Write the action taken or reactants used in the left column. Write the effects of the action or reactants in the right column.

| Action or Reactants | Reaction |
|---|---|
|  |  |

**68** © 2005 Walch Publishing

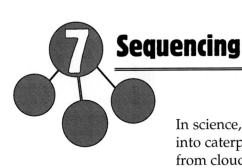

# Sequencing

In science, things happen in a particular order. Butterflies lay eggs, which hatch into caterpillars, which develop pupae, from which adults emerge. Stars develop from clouds of gas, fuse hydrogen for most of their lives, and then go supernova. Knowing the sequence of a series of events helps you understand what you are looking at when you can see only one part of the sequence. You know what came before and what is likely to happen in the future.

For example, imagine that you watch two birds constructing a nest. If you already know the sequence of their natural year, you would know that the building of the nest came after courtship behavior. You would know that egg laying and hatching, and caring for young chicks would follow. You would understand that at some point the chicks would begin to fly, leaving the nest empty through the winter.

Sequencing is useful in all content areas. However, it has many useful applications in science. For example, you might use sequence to organize observations in the wild, to chart the development of an organism, or to show the evolution of a class of organisms over time. You might use it to show the progress of an experiment or to detail the history of a scientific movement.

There are many ways you can show the order of events in science. If the events are recurring, you may show them in a cycle. If the events come from specific causes, you might show them in a cause and effect map. Experiments can be shown using the scientific method.

We have already seen several types of graphic organizers that can be used in sequencing. In this section, we will look at one specific type: the time line. Time lines are used in science, history, social studies, and language arts.

© 2005 Walch Publishing Content-Area Graphic Organizers: Science

**Time Lines**  The most well-known sequencing graphic organizer is a time line. The role of a time line is to show a time-based relationship among events, species, or other things.

**Using Time Lines**  Use a time line when you want to show

- events in a given discipline over time

- when various species lived

- events that may be extra-historical, such as geologic ages or the evolution of the universe

Creating a time line is a simple process once you have the data. You have probably created time lines for other subject areas. In science, the most important thing is to make sure that the space that separates events on your time line is representative of the length of time between the events.

To make a time line, define your beginning and ending times. Next, mark off your time line so that the time intervals and space intervals make sense, or clearly mark beginnings and endings of intervals. Finally, write in the dates, epochs, events, species names, and so forth, that you have collected.

**Time Lines in Action**  Read the article below about geologic ages. Then look at the time line that follows to see how the information has been arranged on the time line.

## Earth's Geologic Ages

Through Earth's history, there have been four great eras. The first—and longest—is called the Precambrian Era. This includes any time before about 570 million years ago (mya). During this era, Earth was created by a bombardment of debris from outer space, the first volcanoes erupted, and the first rain fell, creating the first oceans. At the end of the Precambrian Era, simple life forms arose.

When multicellular life forms arose, around 570 mya, a new era began, the Paleozoic Era. Paleozoic means "old life." The first part of this era was the Cambrian Period, from 570–505 mya. Tiny invertebrates arose during this time. Next came the Ordovician Period. This period lasted from 505–438 mya. During this time, invertebrates grew large and diversified. Next came the Silurian Period, the age of the fishes. This period lasted from 438 mya to 408 mya. The Devonian Period, from 408–360 mya, saw the first terrestrial plants and life on land. The Carboniferous Period is well-known for its giant fern and conifer forests. These eventually became Earth's coal and oil fields. Invertebrates on the land were gigantic—dragonflies had wingspans of two meters. The Carboniferous period began around 360 mya and ended around 286 mya. The last period in the Paleozoic Era was the Permian Period. It began around 286 mya and ended around 245 mya. During this period, vertebrates on land were common—both amphibians and reptiles. They included the animals that would one day become mammals.

At the end of the Permian Period, a great mass extinction occurred. This ended the Paleozoic Era and began the Mesozoic Era, or the "middle life" era. This era is often called the Age of the Dinosaurs, because all three periods had dinosaur life. Dinosaurs were small during the Triassic Period, which began 245 mya and ended 208 mya. However, in the Jurassic Period (208–144 mya), dinosaurs became gigantic. The largest dinosaurs that ever lived on Earth lived during the Jurassic Period. Brachiosaurus and Apatosaurus were large herbivores that lived during this period. The Cretaceous Period (144–65 mya) had large, fierce predators, such as Tyrannosaurus.

The last great era of life is the Cenozoic Era, or "modern life" era. The first period, the Tertiary, began 65 mya and ended around 54 mya. It is known as the Age of Mammals. Gigantic mammals arose. The modern period is the Quaternary Period. It has continued from 54 mya to the present day and includes all modern animals, including humans.

| Precambrian < 570 mya* | Paleozoic 570 mya–245 mya | Mesozoic 245 mya–65 mya | Cenozoic 65 mya–present day |
|---|---|---|---|
| | Cambrian 570–505 mya Ordovician 505–438 mya Silurian 438–408 mya Devonian 408–360 mya Carboniferous 360–286 mya Permian 286–245 mya | Triassic 245–208 mya Jurassic 208–144 mya Cretaceous 144–65 mya | Tertiary 65–54 mya Quaternary 54 mya–present day |

* million years ago

© 2005 Walch Publishing

Use the essay below to create a time line on page 73. As you read, consider what your beginning and ending times will be and where the other events fall in between.

## The Evolution of the Horse

Horses evolved over a long period of time during the Cenozoic era. They are special fossils, because they provide what appears to be a complete fossil record.

The very first horse was called *Hyracotherium,* and it lived in the early Eocene epoch of the Tertiary period. It was very small, about half a meter high at the shoulder, and looked more like a dog than a horse. It was once called *Eohippus,* which means "dawn horse." Although this little animal did not resemble modern horses, it was almost certainly the ancestor of the modern horse. Like modern horses, it had grinding molars and probably ate fruit and some grasses. This creature survived for most of the Eocene (about 20 million years) with only minor evolutionary changes.

Next arose a close relative—*Orohippus,* which looked a lot like *Hyracotherium,* but with stronger grinding molars.

*Epihippus* came from *Orohippus* in the middle of the Eocene epoch. These fossils showed even greater strengthening of the molars, as the little horses' diets changed from soft plant material to tough grass.

*Mesohippus celer* appeared suddenly in the late Eocene, approximately 40 million years ago. It was slightly larger than its early ancestors and didn't look quite so much like a dog.

The next horse, *Miohippus,* was distinctly larger. It was the size of a small deer, and it began to look more like a modern horse. *Mesohippus* finally died out in the mid-Oligocene. *Miohippus* continued for a while as it was, and then, in the early Miocene (24 mya) it began to change into several species fairly rapidly. The horse family began to split into at least two main branches, and a smaller line that soon died out.

The large branches included three-toed horses called *anchitheres.* These animals lived for tens of millions of years. The other branch moved from browsing on low bushes to grazing in open grasslands. A third branch, the pygmy horses, did not long survive.

The grazing horses, however, survived very well. They became the horses of the late Tertiary and Quaternary periods. They also became very fast runners. In order of their appearance, they were: *Kalobatippus, Parahippus, Merychippus, Pliohippus, Astrohippus, Dinohippus,* and finally, modern-day *Equus.* Each of these animals was larger than the one before it.

**Time Line**  Use the information in the essay to complete the time line. It has been placed vertically to give you more room.

## Then and Now Comparison

There are many types of sequencing organizers. Here is another organizer you could use to show changes over time. Write the previous situation or condition under the THEN label, and write the current situation or condition under the NOW label.

| Then | Now |
| --- | --- |
|  |  |

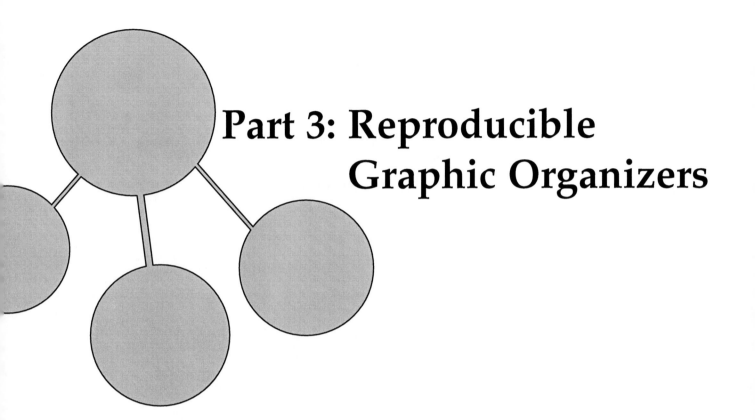

# Part 3: Reproducible Graphic Organizers

# Web

Write the main idea in the center oval. Write the information about the topic in the other ovals.
Add or delete lines and ovals as needed.

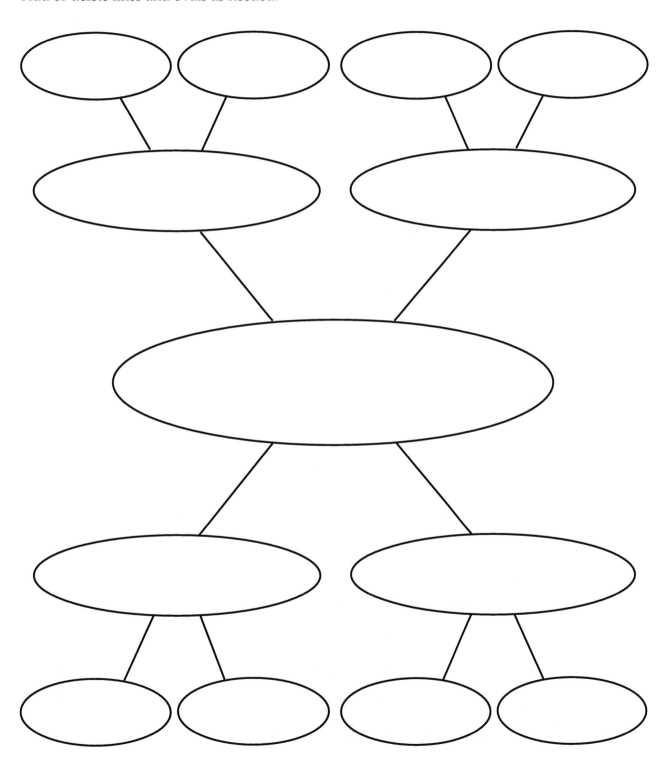

# Table

Write the items being listed at the start of each row. Write the categories of information at the top of each column. Add or delete rows and columns as needed.

| | | | | |
|---|---|---|---|---|
| | | | | |
| | | | | |
| | | | | |
| | | | | |
| | | | | |

# Problem and Solution Chart

Write the problem in the horizontal box at the top. Write possible solutions in the vertical boxes. Once you have investigated the possible solutions, write the most successful one in the horizontal box at the bottom.

**Problem**

**Possible Solution 1**          **Possible Solution 2**          **Possible Solution 3**

**Best Solution**

Write the root of the hierarchy in the box at the top. Write the next level in the hierarchy in the boxes below the top box. Continue until you have added all the levels in the hierarchy. Add or delete lines and boxes as needed.

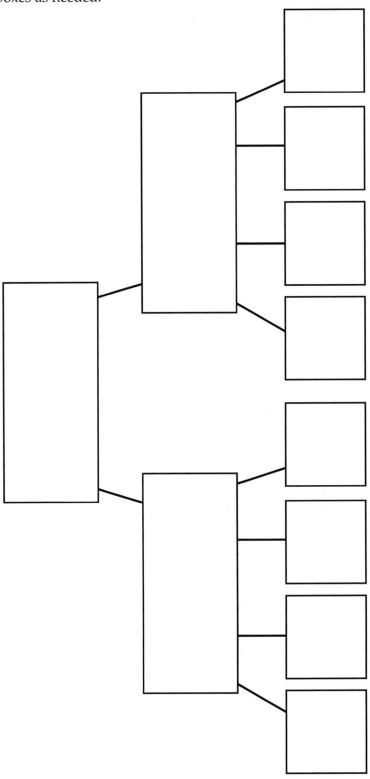

**80**

© 2005 Walch Publishing

# Cladogram

Write the earliest organism at the base of the diagram. Add nodes and branches as new species diverge.

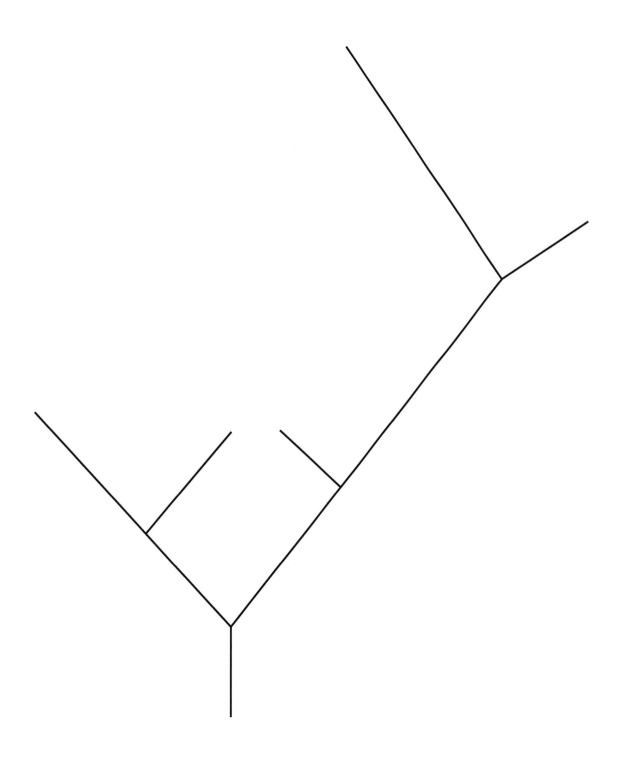

Write the limits at the ends of the scale. Add interal markings if needed. Then write each item in the appropriate place on the continuum.

# Cycle

Write the important stages of the cycle in the boxes. Add or delete boxes as needed.

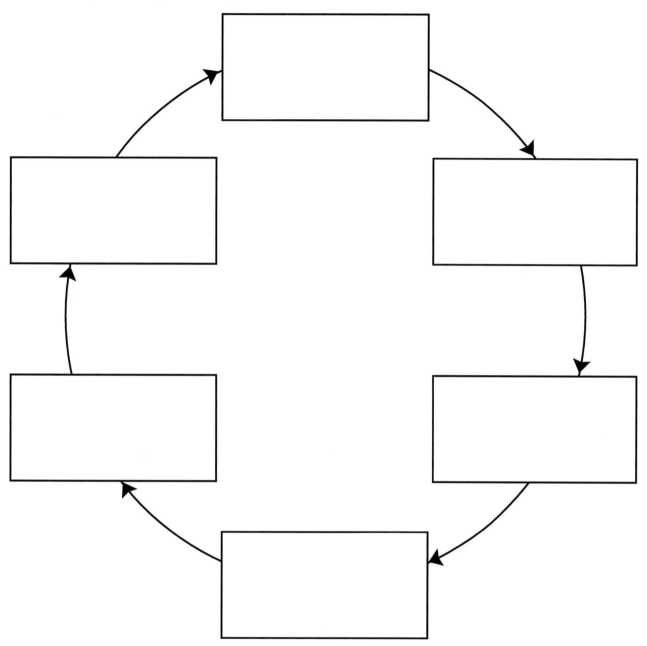

# Free-body Diagrams

Use the centimeter grid below to create a free-body diagram.

© 2005 Walch Publishing

# Pyramid Chart

Write the subgroup with the greatest number of members in the lowest section of the chart. Write each successively smaller subgroup in the upper sections. Add or delete sections as needed.

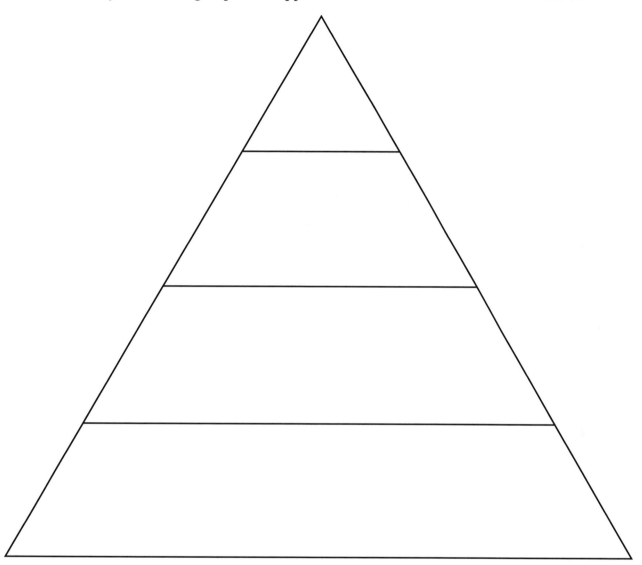

# Comparison Matrix

Write the characteristics to be compared across the top. Write the items to be compared at the beginning of each row. Add or delete rows and columns as needed. Then put check marks in the boxes where rows and columns meet to show that an item has a certain characteristic.

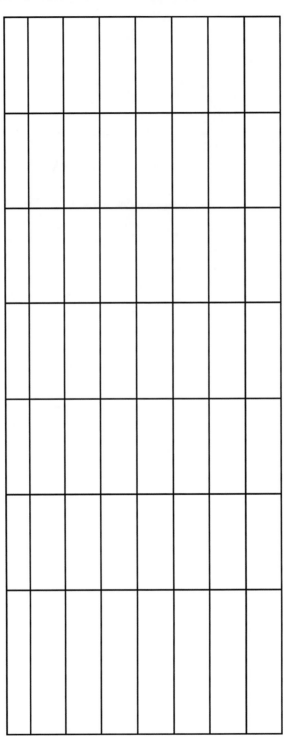

Write the names of the items to be compared in the circles. Write shared attributes in the overlapping area. Write unique attributes in the areas that do not overlap.

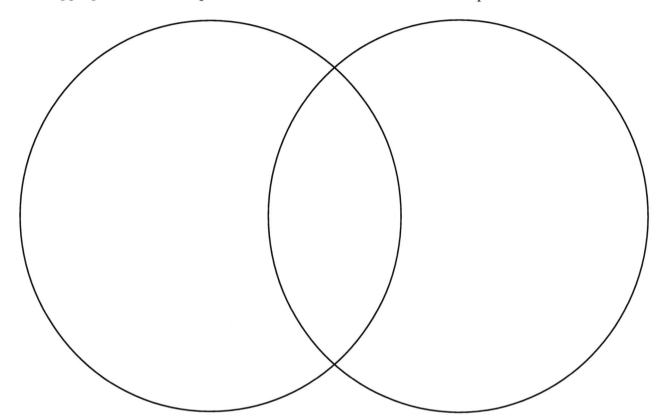

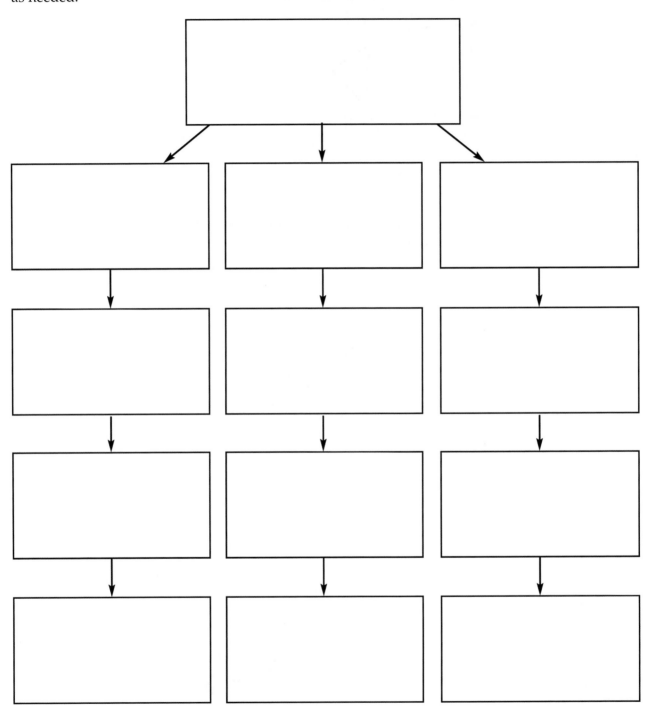

# Multiple-Effect Map

Write the cause in the box at the top. Write the effects in the boxes below. Add or delete boxes as needed.

# Fishbone Map

Write the effect in the box. Write the causes on the branching lines. Add or delete lines as needed.

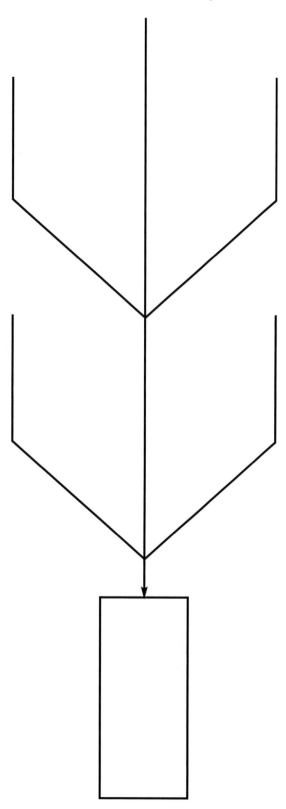

# Jellyfish Map

Write the cause in the horizontal box at the top. Write the effects in the vertical boxes. Add or delete boxes as needed.

**90**

© 2005 Walch Publishing

Write the question you want to explore in the box at the top. Then fill in the other boxes as you complete each step.

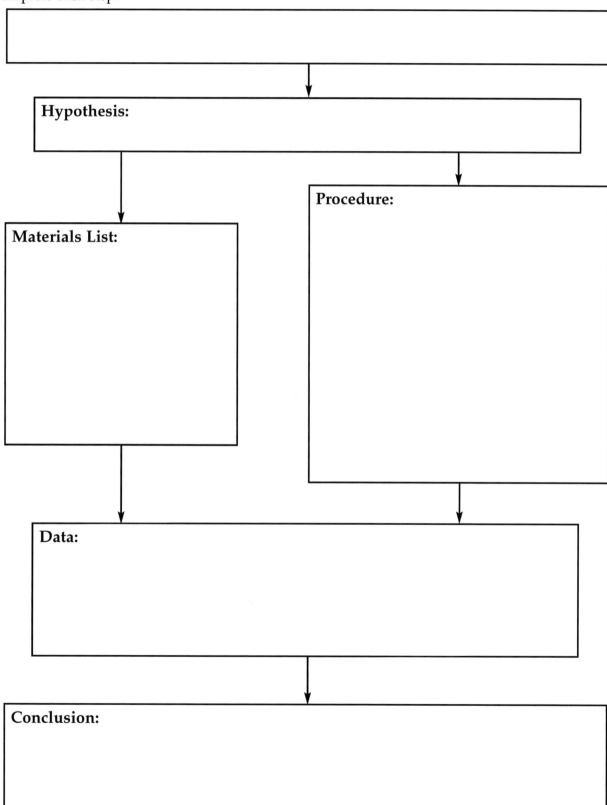

Hypothesis:

Materials List:

Procedure:

Data:

Conclusion:

# Uncertainty Chart

Write the uncertainty factors in the boxes at the top. Find the total uncertainty factor, then use it as a multiplier to find the margin for error and experiment assurance.

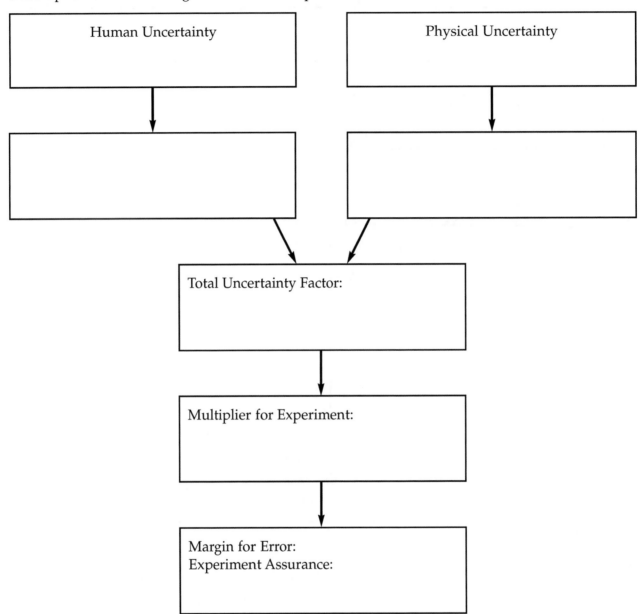

Human Uncertainty

Physical Uncertainty

Total Uncertainty Factor:

Multiplier for Experiment:

Margin for Error:
Experiment Assurance:

**92**
© 2005 Walch Publishing

Write the action taken or reactants used in the left column. Write the effects or reactions in the right column.

| Action or Reactants | Reaction |
|---|---|
| | |

# Time Line

Make the start and end of the time range at the ends of the time line. Mark off intervals on the line. Then add the dates or events that you wish to show.

**94**

© 2005 Walch Publishing

# Then and Now Comparison

Write the previous situation or condition under the THEN label, and write the current situation or condition under the NOW label.

| Then | Now |
|------|-----|
|      |     |

**95**

# Answer Key

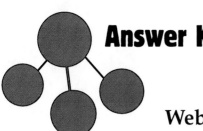

# Answer Key: Lesson 2

## Web, page 11

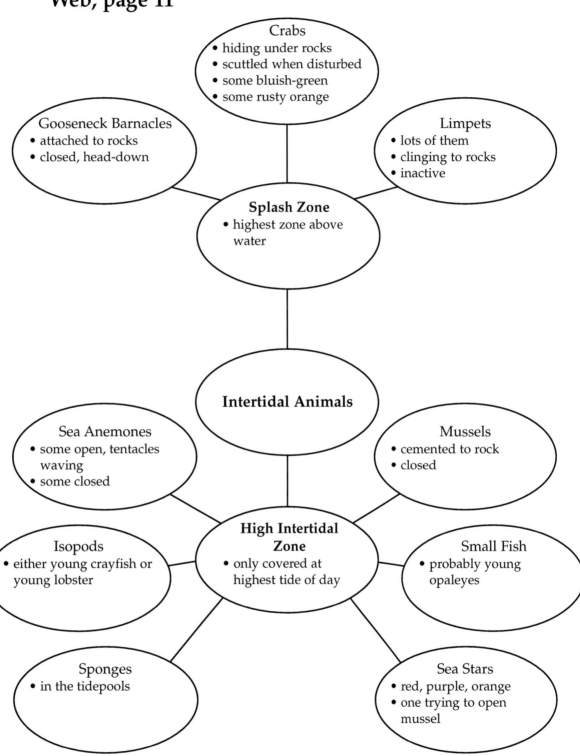

**Crabs**
- hiding under rocks
- scuttled when disturbed
- some bluish-green
- some rusty orange

**Gooseneck Barnacles**
- attached to rocks
- closed, head-down

**Limpets**
- lots of them
- clinging to rocks
- inactive

**Splash Zone**
- highest zone above water

**Intertidal Animals**

**Sea Anemones**
- some open, tentacles waving
- some closed

**Mussels**
- cemented to rock
- closed

**Isopods**
- either young crayfish or young lobster

**High Intertidal Zone**
- only covered at highest tide of day

**Small Fish**
- probably young opaleyes

**Sponges**
- in the tidepools

**Sea Stars**
- red, purple, orange
- one trying to open mussel

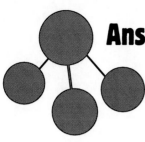

# Answer Key: Lesson 2

## Table, page 15

Answers will vary. Sample answer:

| Chemicals Combined | Exothermic or Endothermic? | Gave off Gas? | Changed Color? | Precipitated? |
|---|---|---|---|---|
| sodium acetate solution, sodium acetate crystal | got very warm—exothermic | no | no | made crystals |
| barium hydroxide, ammonium nitrate | got very cold | no | no | no |
| acetic acid and water, phenolphthalein | no | no | turned bright pink | no |
| solution from Experiment 3, ammonia, water | neither | no | color vanished | no |
| acetic acid, sodium bicarbonate | slight warming—exothermic | bubbles of carbon dioxide | no | no |
| lead nitrate, ammonia | slight cooling—endothermic | creamy soaplike gas | changed from colorless to white | small bits of lead |

# Answer Key: Lesson 3

## Hierarchical Diagram, page 21

Answers will vary. Sample answer:

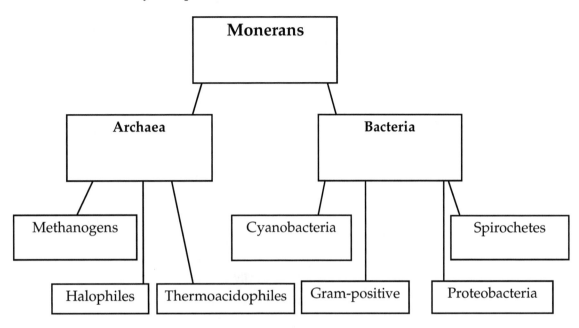

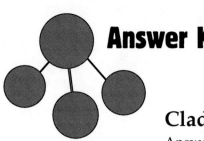

# Answer Key: Lesson 3

## Cladogram, page 26

Answers will vary. Sample answer:

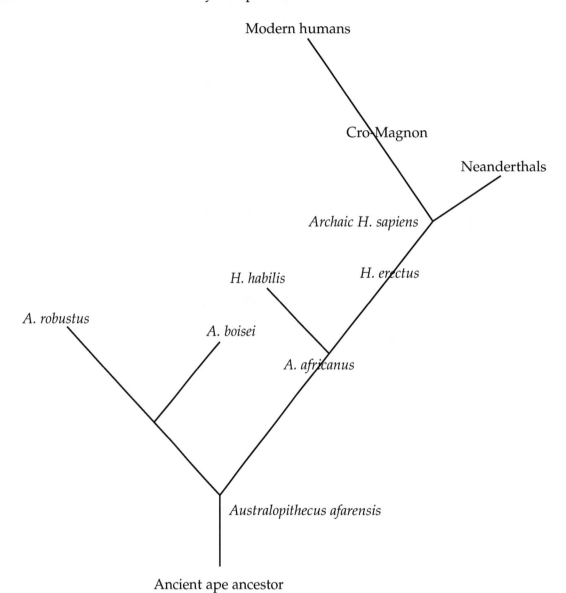

Modern humans

Cro-Magnon

Neanderthals

*Archaic H. sapiens*

*H. habilis*          *H. erectus*

*A. robustus*

*A. boisei*

*A. africanus*

*Australopithecus afarensis*

Ancient ape ancestor

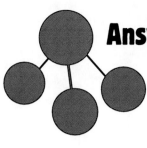

# Answer Key: Lesson 3

## Continuum Scale, page 30

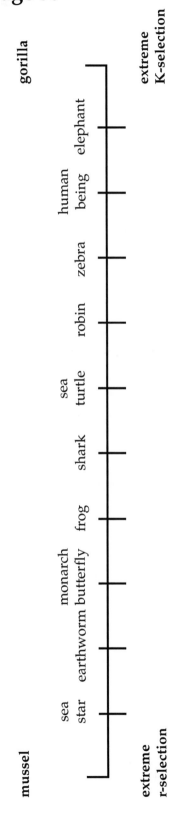

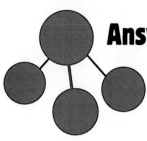

## Cycle Diagram, page 34

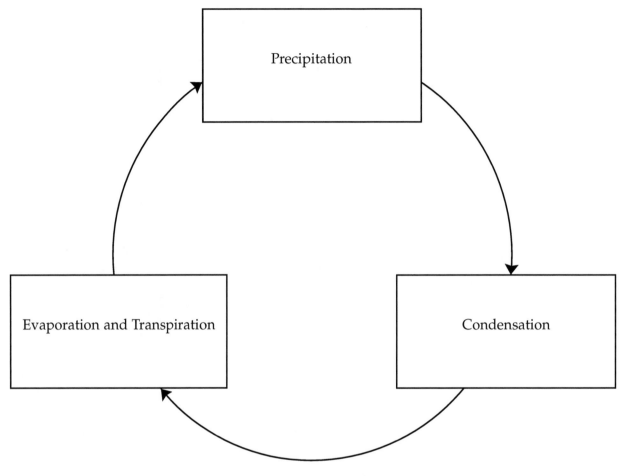

**The Water Cycle**

 © 2005 Walch Publishing

# Answer Key: Lesson 3

## Free-body Diagram, page 38

Answers will vary. Sample answer:

girls ————————— ← Applied force | ————————— boys

**First Day**

girls ————————— | Applied force → ————————— boys

**Second Day**

girls ————————— | Applied force ————————→ boys

**Third Day**

girls ← ————————— Applied force | ————————— boys

**Fourth Day**

girls ————— ← Applied force | ————————— boys

**Fifth Day**

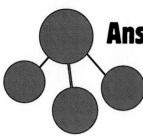

# Answer Key: Lesson 4

## Comparison Matrix, page 45

| Mineral | Cubic | Hexagonal | Tetragonal | Trigonal | Orthorhombic | Monoclinic | Triclinic |
|---------|-------|-----------|------------|----------|--------------|------------|-----------|
| Sodium chloride | ✔ | ✗ | ✗ | ✗ | ✗ | ✗ | ✗ |
| Graphite | ✗ | ✔ | ✗ | ✗ | ✗ | ✗ | ✗ |
| Cat's-eye quartz | ✗ | ✗ | ✔ | ✗ | ✗ | ✗ | ✗ |
| Quartz | ✗ | ✗ | ✗ | ✔ | ✗ | ✗ | ✗ |
| Sulfur | ✗ | ✗ | ✗ | ✗ | ✔ | ✗ | ✗ |
| Gypsum | ✗ | ✗ | ✗ | ✗ | ✗ | ✔ | ✗ |
| Turquoise | ✗ | ✗ | ✗ | ✗ | ✗ | ✗ | ✔ |

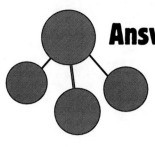

## Multiple-Effect Map, page 53

Answers will vary. Sample answer:

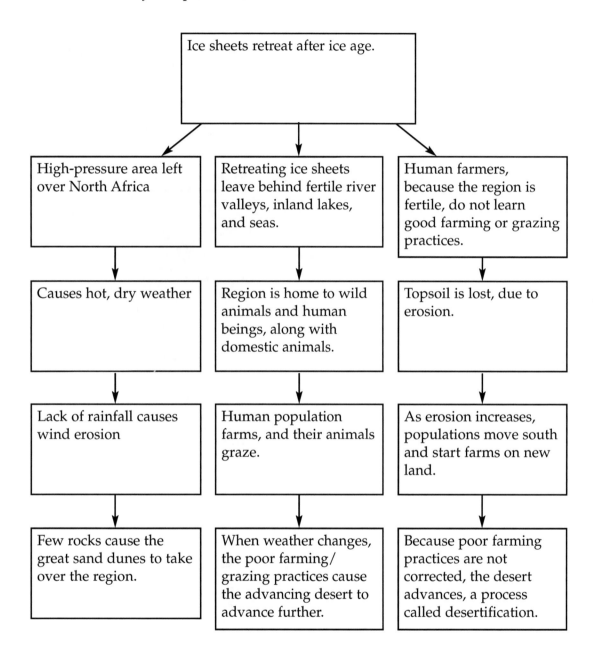

Ice sheets retreat after ice age.

| High-pressure area left over North Africa | Retreating ice sheets leave behind fertile river valleys, inland lakes, and seas. | Human farmers, because the region is fertile, do not learn good farming or grazing practices. |
| --- | --- | --- |
| Causes hot, dry weather | Region is home to wild animals and human beings, along with domestic animals. | Topsoil is lost, due to erosion. |
| Lack of rainfall causes wind erosion | Human population farms, and their animals graze. | As erosion increases, populations move south and start farms on new land. |
| Few rocks cause the great sand dunes to take over the region. | When weather changes, the poor farming/ grazing practices cause the advancing desert to advance further. | Because poor farming practices are not corrected, the desert advances, a process called desertification. |

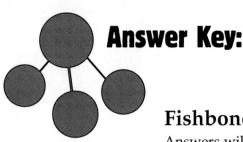

# Answer Key: Lesson 5

## Fishbone Map, page 57
Answers will vary. Sample answer:

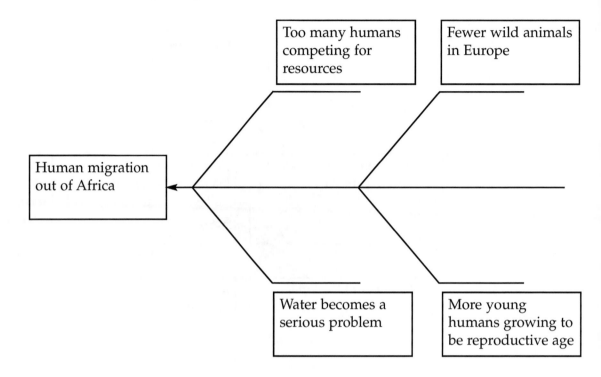

Too many humans competing for resources

Fewer wild animals in Europe

Human migration out of Africa

Water becomes a serious problem

More young humans growing to be reproductive age

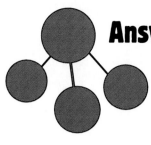

# Answer Key: Lesson 6

## The Scientific Method, page 63

Answers will vary. Sample answer:

Is it necessary to have oxygen in a tank for cyanobacteria to grow?

**Hypothesis:** While cyanobacteria produces oxygen, it does not require it to grow.

**Materials List:**

1. Cyanobacteria (2 strands)
2. 2 beakers
3. Plastic wrap
4. Goldfish
5. Straw
6. Ruler marked with mm

**Procedure:**

1. Let water in one beaker stand until all bubbles are gone. Keep fish in water for a day, then put it back in its tank. Remove excess surface oxygen with a straw.

2. Measure two strands of cyanobacteria at regular intervals— one in oxygen reduced water, one in freshwater.

3. At the end of five days, check to see which is longer.

**Data:**
Day 1: Oxygen reduced: 13 mm    Oxygen rich: 13 mm
Day 2: Oxygen reduced: 15.5 mm    Oxygen rich: 15.5 mm
Day 3: Oxygen reduced: 17 mm    Oxygen rich: 17.1 mm
Day 4: Oxygen reduced: 17.5 mm    Oxygen rich: 17.4 mm
Day 5: Oxygen reduced: 18.1 mm    Oxygen rich: 18.1 mm

**Conclusion:** It is not necessary to have oxygen to grow cyanobacteria.

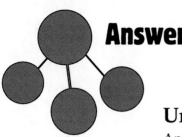

# Answer Key: Lesson 6

## Uncertainty Chart, page 67

Answers will vary. Sample answer:

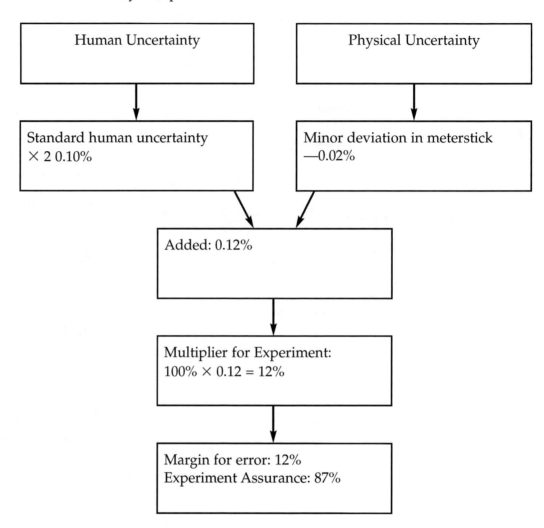

```
┌─────────────────────────┐        ┌─────────────────────────┐
│   Human Uncertainty     │        │   Physical Uncertainty  │
└───────────┬─────────────┘        └───────────┬─────────────┘
            ↓                                  ↓
┌─────────────────────────┐        ┌─────────────────────────┐
│ Standard human          │        │ Minor deviation in       │
│ uncertainty × 2 0.10%   │        │ meterstick —0.02%        │
└───────────┬─────────────┘        └───────────┬─────────────┘
            └──────────┐            ┌───────────┘
                       ↓            ↓
            ┌─────────────────────────────────┐
            │ Added: 0.12%                    │
            └────────────────┬────────────────┘
                             ↓
            ┌─────────────────────────────────┐
            │ Multiplier for Experiment:      │
            │ 100% × 0.12 = 12%               │
            └────────────────┬────────────────┘
                             ↓
            ┌─────────────────────────────────┐
            │ Margin for error: 12%           │
            │ Experiment Assurance: 87%       │
            └─────────────────────────────────┘
```

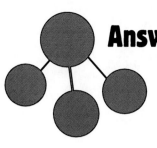

# Answer Key: Lesson 7

## Time Line, page 73

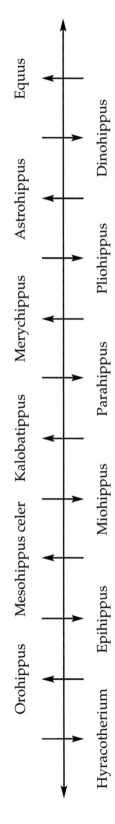

Equus

Dinohippus

Astrohippus

Pliohippus

Merychippus

Parahippus

Kalobatippus

Miohippus

Mesohippus celer

Epihippus

Orohippus

Hyracotherium

# Share Your Bright Ideas

## We want to hear from you!

Your name_____Date_____

School name_____

School address_____

City _____State _____Zip_____Phone number (_____)_____

Grade level(s) taught_____Subject area(s) taught_____

Where did you purchase this publication?_____

In what month do you purchase a majority of your supplements?_____

What moneys were used to purchase this product?

____School supplemental budget ____Federal/state funding ____Personal

**Please "grade" this Walch publication in the following areas:**

Quality of service you received when purchasing ...................................................... A    B    C    D

Ease of use................................................................................................................. A    B    C    D

Quality of content...................................................................................................... A    B    C    D

Page layout ............................................................................................................... A    B    C    D

Organization of material ........................................................................................... A    B    C    D

Suitability for grade level ......................................................................................... A    B    C    D

Instructional value.................................................................................................... A    B    C    D

COMMENTS:_____

_____

What specific supplemental materials would help you meet your current—or future—instructional needs?

_____

Have you used other Walch publications? If so, which ones?_____

May we use your comments in upcoming communications?      ____Yes      ____No

Please **FAX** this completed form to **888-991-5755**, or mail it to

**Customer Service, Walch Publishing, P. O. Box 658, Portland, ME 04104-0658**

We will send you a **FREE GIFT** in appreciation of your feedback. **THANK YOU!**